KT-440-460

Contents

Mike McGrath

Building Android Apps

In easy steps is an imprint of In Easy Steps Limited
4 Chapel Court · 42 Holly Walk · Leamington Spa
Warwickshire · United Kingdom · CV32 4YS
www.ineasysteps.com

Notice of Liability
Every effort has been made to ensure that this book contains accurate
and current information. However, In Easy Steps Limited and the
author shall not be liable for any loss or damage suffered by readers
as a result of any information contained herein.

Trademarks
All trademarks are acknowledged as belonging to their respective
companies.

In Easy Steps Limited supports The Forest Stewardship Council (FSC),
the leading international forest certification organization. All our titles
that are printed on Greenpeace approved FSC certified paper carry the
FSC logo.

MIX
Paper from
responsible sources
FSC® C020837

Printed and bound in the United Kingdom

ISBN 978-1-84078-528-9

Foreword

The creation of this book has provided me, Mike McGrath, a welcome opportunity to demonstrate how App Inventor For Android can be used to easily create applications for today's portable devices. All examples I have given in this book demonstrate app features using the current online version of App Inventor hosted by the MIT Center for Mobile Learning.

Conventions in this book

Hello.apk

The examples provide screenshots of the actual App Inventor blocks used to implement each step for clarity. Additionally, in order to identify each example described in the steps a colored icon and an Android Application Package file name appears in the margin alongside the steps.

Grabbing the source code

For convenience I have placed source code files from the examples featured in this book into a single ZIP archive. You can obtain the complete archive by following these easy steps:

1. Browse to **http://www.ineasysteps.com** then navigate to the "Resource Center" and choose the "Downloads" section

2. Find "Building Android Apps in easy steps" in the "Source Code" list, then click on the hyperlink entitled "All Code Examples" to download the complete ZIP archive

3. Now extract the archive contents to any convenient location on your computer – the source code for each example is contained in its own individual ZIP archive

4. You may then upload any example into App Inventor using the More Actions | Upload Source menu on the My Projects view to select a source code ZIP archive

I sincerely hope you enjoy discovering the powerful exciting possibilities of App Inventor For Android and have as much fun with it as I did in writing this book.

Mike McGrath

1 Getting started

Welcome to the exciting world of application development for Android. This chapter shows how to establish an app development environment and demonstrates how to create a simple Android app.

Introducing Android

Android is an operating system for mobile/cell phones and tablets, in much the same way that Microsoft Windows is an operating system for PCs. The Android operating system is maintained by Google and comes with a range of useful features as standard.

Standard Android features include Google Search and Google Maps, which means you can easily search for information on the web and find directions from your phone – as you would on your computer. This is handy for discovering things like train times and getting directions when out and about. Other Google services, such as Gmail and Google Earth can also be accessed from cell phones running the Android operating system. You can easily check Facebook and Twitter profiles too, through a variety of applications (apps) – making it ideal for social networking.

There is a huge range of custom apps available to download from the Google Play Shop. For example there are camera apps such as "Camera 360" – that allow you to take photos with artistic effects, and music player apps such as "Winamp" – that allow you to import MP3s and create playlists, and popular game apps, such as "Angry Birds" – that provide great fun entertainment.

Don't forget

Many apps available on the Android Market are free but some will require payment.

Android is an open-source operating system, built on the open-source Linux Kernel, which means it can be easily extended to incorporate new cutting-edge technologies as they emerge. Android was brilliantly designed, from the ground-up, to enable developers to create compelling apps that can fully exploit all the host device's capabilities. For example, an app can access all of a phone's core functionality such as making calls, sending text messages, or taking photos. The Android platform will continue to evolve as the developer community works together to build innovative mobile applications and you can be part of this exciting innovation process with App Inventor for Android.

App Inventor for Android

App Inventor enables you to develop applications for Android phones using a web browser and either a connected phone or emulator – without writing a single line of code. It is a web-based tool developed by Google Labs in which the App Inventor servers store your work and help you keep track of your projects. When the app is finished you can package it to produce an "application package" (Android **.apk** file) that can be shared around and installed on any Android phone, just like any other Android app. App Inventor is supported by a wide range of operating systems and web browsers, with these minimum specifications:

Computer operating system requirements:

- **Windows** : Windows XP, Windows Vista, Windows 7+

- **Linux** : Ubuntu 8+, Debian 5+

- **Mac** : Mac OS X 10.5+

Web browser requirements:

- **Internet Explorer** : 7.0+

- **Mozilla Firefox** : 3.6+

- **Google Chrome** : 4.0+

- **Apple Safari** : 5.0+

Additionally App Inventor requires two pieces of software to be installed on your computer:

- **Java** : to provide software libraries for the web-based App Inventor development environment

- **App Inventor Setup** : to provide client-side support for the web-based App Inventor development environment

The installation process for both Java and App Inventor Setup is described on the following pages.

Hot tip

Since its original development at Google Labs, App Inventor is now hosted by the MIT Center for Mobile Learning (CML).

Installing Java

In order to use the App Inventor development environment your computer needs to provide Java support libraries. Follow these easy steps to install Java on your own system:

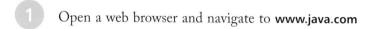 **1** Open a web browser and navigate to **www.java.com**

 **2** Click the "Free Java Download" button to start the Java installation process

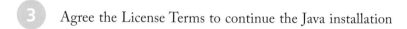 **3** Agree the License Terms to continue the Java installation

4 When the Java installation completes, navigate to the Test Page at **www.java.com/en/download/testjava.jsp** to verify the Java configuration on your computer – a notice like the one below should then appear to announce that your system is correctly configured for Java

Your Java is working

Your Java configuration is as follows:

Vendor: Oracle Corporation
Version: Java SE 7
Operating System: Windows 7 6.1
Architecture: x86

 Beware

App Inventor is not designed for OpenJDK – use only the Java packages from **java.com**.

5 Now navigate to the App Inventor Java Test Page at **appinventor.mit.edu/learn/setup/misc/JWSTest/ApplnvJWSTest.html** to check that your browser is properly configured for Java – a notice like the one opposite should then appear to announce that your web browser is correctly configured for Java

Check Java on your computer for App Inventor

Part 1: App Inventor is checking your browser for running programs with Java Web Start. *Checking ...*

✓ Your browser appears to be configured properly

6 Finally click the "Launch" button on the App Inventor Java Test Page to check that your computer can launch applications with Java Web Start – a Java-based Notepad application should then appear to indicate that your system is correctly configured for Java Web Start

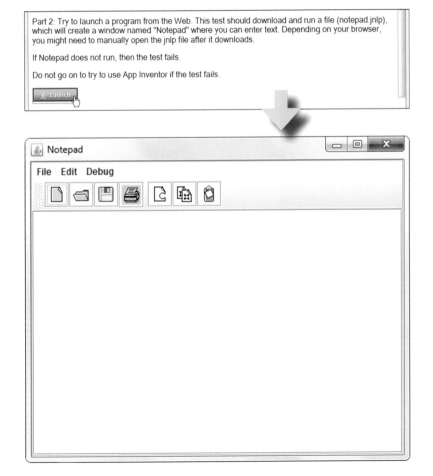

Part 2: Try to launch a program from the Web. This test should download and run a file (notepad.jnlp), which will create a window named "Notepad" where you can enter text. Depending on your browser, you might need to manually open the jnlp file after it downloads.

If Notepad does not run, then the test fails.

Do not go on to try to use App Inventor if the test fails.

Launch

Notepad

File Edit Debug

Don't forget

All Java configuration tests must succeed before you can use App Inventor – use the suggestions on the App Inventor Test Page to correct any failures before proceeding.

Installing App Inventor

Having installed the Java libraries, as described on the previous page, your computer now needs to provide client-side support for the App Inventor development environment. Follow these easy steps to implement App Inventor Setup on your own system:

1 Open a web browser and type the App Inventor Setup file's location into the location field, such as
http://dl.google.com/dl/appinventor/installers/windows/ appinventor_setup_installer_v_1_2.exe

2 Now hit Return to download App Inventor Setup to any convenient system location, such as your Desktop

3 Execute the App Inventor Setup file, such as **AppInventor_Setup_Installer_v_1_2.exe** (~92 MB) to start the installation process

4 Click through the steps of the installer to continue App Inventor Setup – being sure not to change the suggested default installation location on your system

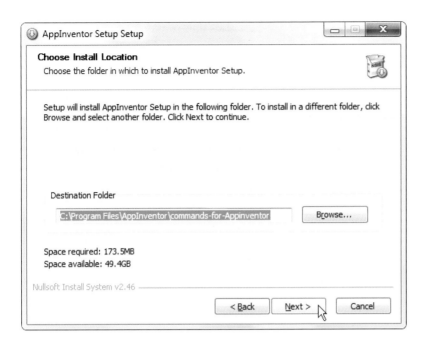

Hot tip

The typical App Inventor Setup default location is **C:\Program Files\ Appinventor\commands- for-Appinventor**.

App Inventor provides a device emulator for app development without a physical device but App Inventor Setup installs the Google USB driver that supports many common Android phones. Other devices will require you to obtain and install a USB driver from the device manufacturer

5   To test driver installation for an Android phone first connect the phone to your computer via a USB socket

6   Now open a Command Prompt window in the App Inventor directory, then enter the command **adb devices** – a device number should appear in the "List of devices attached", similar to the one below, and you're good to go

Don't forget

USB driver information links for many manufacturers can be found online at **developer.android.com/ sdk/oem-usb.html**.

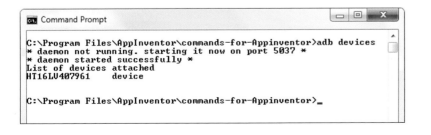

Beginning your first app

Having installed the Java libraries and implemented the App Inventor Setup, as described on the previous pages, you are ready to begin developing your first application for the Android platform – that will generate the traditional first program greeting:

 Open a web browser and navigate to the App Inventor at **appinventor.mit.edu** then accept the "Terms of Service" if requested to open the App Inventor "Projects" page

Upon your first visit the App Inventor Projects page appears with no previously created projects and looks like this:

Hot tip

You may be prompted to sign in to a Google Account before being allowed access to the App Inventor – sign in with an existing account or create a new account in order to continue.

14

 Click the "New" button to launch the New Android Project dialog, then type the Project Name **Hello** and click OK – to see a new empty project appear

Hot tip

Name new projects capitalized with an uppercase first character.

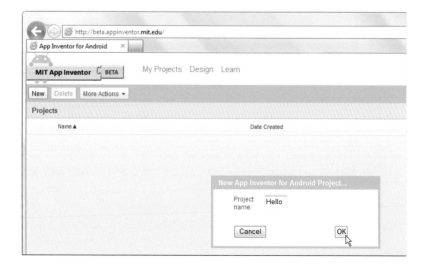

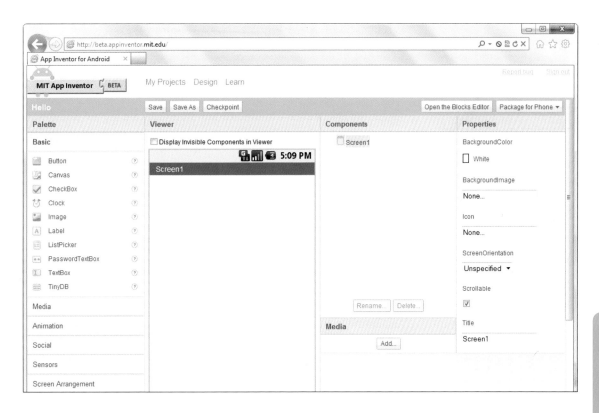

A newly created empty project first appears in App Inventor's "Designer" window. The Designer is where you add components to the application to design the user interface. It contains these five columns, which you should familiarize yourself with:

- **Palette :** containing individual components that can be added to the interface, grouped by category – such as "Basic"

- **Viewer :** visually representing components added to the interface – a screen container component is provided by default

- **Components :** listing components added to the interface – arranged hierarchically beneath the screen component

- **Media :** listing media resources used by the app – such as images, audio, and video files

- **Properties :** listing editable characteristics of the component currently selected in the Viewer column – such as "Screen1"

Don't forget

The Media column is located directly below the Components column.

15

Adding components

The first stage in creating an app is to design the user interface by adding components from the Palette in App Inventor Designer. Follow these easy steps to add components to the Hello project, begun on the previous page, on your own system:

 Click on the Palette column's Basic category to reveal the component items it contains

 Next click on a Label component in the Palette column's Basic category and drag a copy of it to the Viewer column – then release the mouse button to drop it there

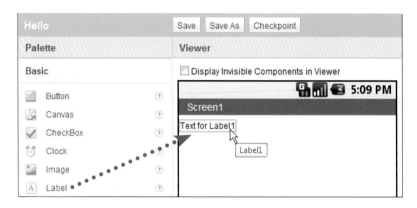

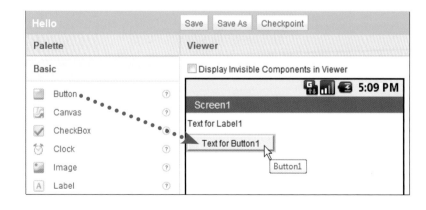 Now click on a Button component in the Palette column's Basic category and drag a copy of it to the Viewer column – then release the mouse button to drop it there

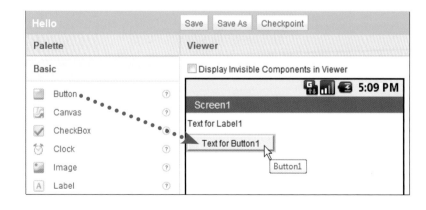

16

4 Click on the Label component added to the Viewer to select it – see its characteristics appear in the Properties column

Items listed in the Properties column can be edited by clicking on them to change text values, check/uncheck boxes, or choose from alternatives that appear in pop-up dialog boxes.

5 Edit the Label component's characteristics in the Properties column to set its BackgroundColor to "Blue", FontBold to checked, Font Size to "30.0", Text to blank, TextColor to "Yellow", and Width to "Fill parent..." – see the Label's appearance change in the Viewer

6 Now click on the Button component in the Viewer to select it – see its characteristics appear in the Properties column

7 Edit the Button component's characteristics in the Properties column to set its Text to "Click Me" – see the Button's appearance change in the Viewer

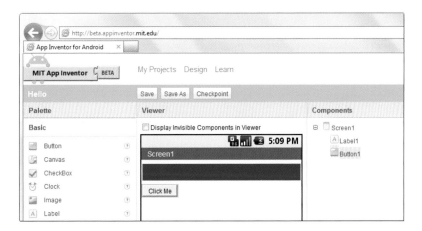

The components added to the Viewer have been automatically named "Label1" and "Button1" by App Inventor but as yet have no functionality. The next stage in creating an app is to add functionality – making the Button respond to a user action by displaying a response message in the Label component.

Adding behavior

The second stage in creating an app is to add functionality to components added to the user interface in App Inventor Designer. This stage employs the App Inventor Blocks Editor that allows you to easily add functionality without writing any program code. Follow these easy steps to add "Click" behavior functionality to components of the Hello project, added on the previous page:

1 On the App Inventor menu bar, click on the "Open the Blocks Editor" button – to launch the Blocks Editor in a new window

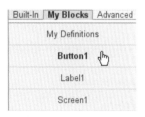

The Blocks Editor has three tabs entitled "Built-In", "My Blocks", and "Advanced", each having a number of uniquely labeled "drawers" – that each contain various individual colored blocks. The blocks contain text and can be snapped together, like jigsaw-puzzle pieces, to form instructions to be performed by the app.

2 Click on the "My Blocks" tab – to reveal its drawers

3 Next click the "Button1" drawer – to reveal its blocks

4 Now drag the top block containing the text "when Button1.Click do" onto the Blocks Editor workspace

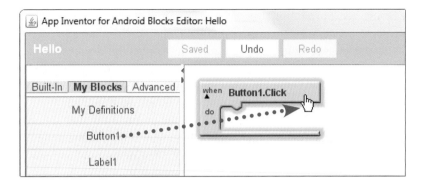

5 Click the "Label1" drawer and drag the "set Label1.Text to" block to the workspace – snapping it into the previous block

Don't forget

Whenever you add a component to the interface in Designer it also gets a drawer added to the My Blocks tab in the Blocks Editor containing blocks to define its behavior.

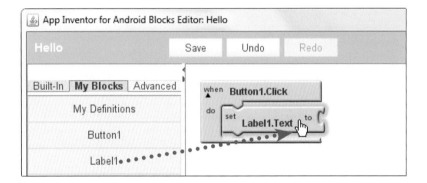

6 Click the Built-In tab, then open the "Text" drawer and snap its "text text" block into the previous block

7 Finally click on the text block's content, to make it editable, and change the content to "Hello World!"

Beware

If you snap together blocks that form an invalid instruction the Blocks Editor will advise you of the error.

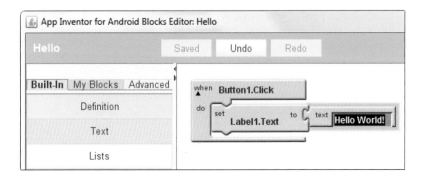

19

Preparing devices

Having added functionality to the components of the Hello project's user interface, on the previous page, their behavior can be tested in App Inventor's virtual Emulator phone device. Follow these easy steps to prepare the phone Emulator:

1 On the Blocks Editor menu bar, click on the "New emulator" button – to start up the phone emulator

A dialog box appears while the phone Emulator is starting to advise you that initialization is underway, but may take a while.

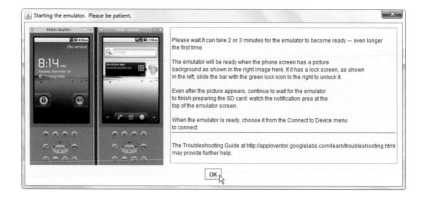

When the Emulator starts you should wait until it displays a colored screen and has finished preparing its SD card.

2 The App Inventor device Emulator opens in a new window depicting a locked phone screen. Click on the green unlock icon and drag it to the right to unlock the device – ready for connection to the App Inventor development environment

A physical Android phone device can also be used to test app behavior if the phone is first placed in "Debug mode". Follow these easy steps to prepare a physical Android phone:

3 Connect an Android phone to a USB socket on your computer, and turn on the phone

4 From the phone's Home screen, tap the Settings button then tap the "Applications" item from the list to bring up the Applications menu

An Android phone will not be able to communicate with App Inventor unless the phone's Debug mode is first enabled.

5 On the Applications menu check the "Unknown sources" item if present, then tap the "Development" item to bring up the Development menu

6 On the Development menu, check both the "USB Debugging" item and the "Stay awake" item

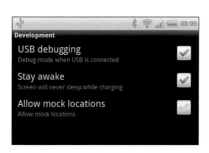
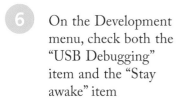

Beware

Set the phone to "Stay awake" as Android stops any running app whenever the screen goes to sleep.

7 Finally tap the Home button to apply the new settings – ready for connection to the App Inventor development environment on your computer.

Running your first app

Having prepared the Android test devices, as described on the previous page, the Hello application can now be executed to see how it performs. Follow these easy steps to run the app on App inventor's virtual phone Emulator:

1. On the Blocks Editor menu bar, click on the "Connect to Device" button – to open a list of connection options

2. Choose the Emulator option to begin connecting App Inventor to the virtual phone emulator

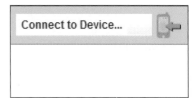

During the connection process the icon to the right of the "Connect to Device" button first changes to flashing yellow, but becomes solid green when the Emulator is connected. The application then gets loaded onto the screen and is ready to run.

3. Click the Button component on the Emulator screen to simulate a user's tap action – see the app implement the behavior defined in the Blocks Editor to apply Label text

Don't forget

Testing an app with the Emulator or a phone is the third stage in creating an Android app – after the Designer and Blocks Editor stages.

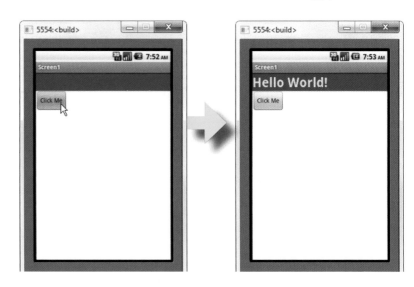

A physical Android phone device can also be used to execute an app behavior to see how it performs. Follow these easy steps to run the Hello app on a physical Android phone:

4 On the Blocks Editor menu bar, click on the "Connect to Device" button – to open a list of connection options

5 Choose the numbered device option to begin connecting App Inventor to the physical Android phone

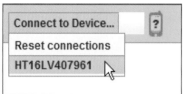

During the connection process the icon to the right of the "Connect to Device" button first changes to flashing yellow, but becomes solid green when the Android phone is connected. The application then gets loaded onto the screen and is ready to run.

6 Tap the Button component on the Android phone screen – see the app implement the behavior defined in the Blocks Editor to apply the Label text greeting message

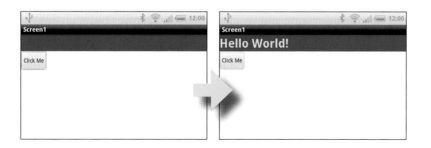

Congratulations, the functionality of your first app has been successfully tested on both the App Inventor virtual phone Emulator and on a connected physical Android phone. Just like any other app, the completed Hello app may now be permanently installed on the Android phone – as described on the next page.

Managing projects

The App Inventor Designer menu bar contains a number of useful buttons for managing your application projects. Completed apps can be packaged into a single compact file for installation on any Android device using the "Package for Phone" button. Follow these steps to package the Hello project from the previous pages:

 Click on the "Package for Phone" button and choose the "Download to this Computer" option from the dropdown menu that appears

App Inventor then packages all the application's source files into a single file named with the project name and a **.apk** file extension – this can then be used to install the app on any Android device.

 When requested save the packaged **Hello.apk** file to any convenient location on your system – such as the Desktop

Hello.apk

Do you want to open or save **Hello.apk** (1.15 MB) from **appinventor.googlelabs.com**? Save ▼

Beware

The Unkown Sources item on the phone's Settings, Applications menu must be checked to allow installation of non-Market applications.

Employ your phone's Application Installer, such as that in HTC Sync, to select the saved packaged file and install the application on the phone – the Hello app now appears alongside all your other installed apps

Whenever you make any change to an application you are developing App Inventor automatically saves the project, ensuring it is constantly up-to-date – so the App Inventor "Save" button is seldom needed. Its "Save As" button is useful to create a new copy of the project, and you then continue working on that new copy. Its "Checkpoint" button, on the other hand, is useful to create a new backup copy of the project, and you then continue working on the original project.

4. Click on the Checkpoint button and accept the suggested default name to create a backup of the Hello project

As you develop more applications with App Inventor each project gets automatically saved in a "My Projects" list. Clickable links at the top of the App Inventor window allow you to easily toggle between the "My Projects" view and the "Design" view at any time. In the My Projects list view projects can be added at any time using the "New" button, and existing projects can be deleted using the "Delete" button.

5. In My Projects check the box against the Hello project backup item, then click the Delete button to remove it

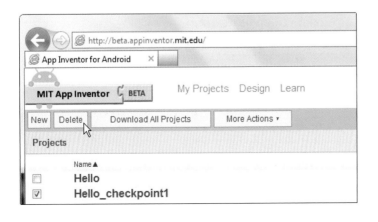

Don't forget

The App Inventor source code for all your projects can be downloaded as a single ZIP archive with the "Download All Projects" button on the My Projects list view.

Summary

- Android is an operating system for mobile/cell phones and tablets

- App Inventor enables you to develop applications for Android devices without writing a single line of code

- Java must be installed on your computer to support the web-based App Inventor development environment

- App Inventor first opens on the Projects page where new projects get started

- The App Inventor Designer is where you add components to the application to design the user interface

- Designer contains Palette, Viewer, Components, Media, and Properties columns

- Interface components can be dragged from the Palette onto the Viewer and their characteristics edited in Properties

- The App Inventor Blocks Editor is where you add functionality to the application to create behaviors

- Blocks Editor contains Built-in, My Blocks, and Advanced tabs that each have a number of drawers

- Blocks can be dragged from the Blocks Editor drawers and snapped together on the workspace to form instructions

- The App Inventor virtual phone Emulator can be used to test application performance

- A physical Android phone device can be used to test application performance

- Test devices are connected to App Inventor by selection from the "Connect to Device" options in Blocks Editor

- Completed apps can be packaged for distribution using the "Package for Phone" button in App Inventor Designer

- Projects are managed from the My Projects list view

2 Designing interfaces

This chapter demonstrates how to use components of the App Inventor Basic Palette for interface design in Android apps.

Enabling buttons

A Button component provides the user an easy way to interact with an Android application. Tapping a button in the interface creates a "Click event" within the app. The application may then respond by calling a Click "event-handler" behavior to perform a task, such as write a response message in a Label component.

Optionally a Button's properties may be changed as the application proceeds to reflect its current status. For example, its "Enabled" property can be toggled on or off – to make the Button active or inactive according to the app requirements.

Button.apk

1 Start a new App Inventor project named "Button", and also set the screen's Title property to "Button"

2 From the Basic Palette, drag one Label component and two Button components onto the Viewer

3 In the Components column, rename the components **lblMessage**, **btnStart** and **btnStop** respectively

4 In the Properties column, edit the components' BackgroundColor, TextColor, FontSize, and Text properties so they resemble the screenshot below:

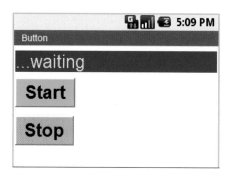

Hot tip

Rename components with an identifying prefix, such as "btn" for Button components – so their component types are easily identifiable in the Components column.

5 Now in in the Properties column, uncheck the **btnStop** component's Enabled property to change its default state – so it will not be active when the application launches

6 Next launch the Blocks Editor – ready to add event-handler behaviors for each Button's Click event

7 From the My Blocks tab, drag a **btnStart.Click** block and a **btnStop.Click** block onto the workspace

8 Next from the My Blocks tab, snap **set lblMessage.Text**, **set btnStart.Enabled**, and **set btnStop.Enabled** blocks into both of the Button's Click blocks on the workspace

Hot tip

The My Blocks tab has drawers for each component that contain the blocks to set and get their property values.

9 From the Built-In tab's Text drawer, snap text blocks into the **set lblMessage.Text** blocks and edit their default text content to assign appropriate Label text values – reflecting a running or stopped state

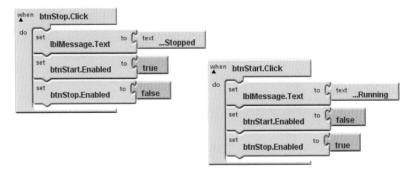

10 From the Built-In tab's Logic drawer, snap **true** and **false** boolean value blocks into the **set btnStart.Enabled**, and **set btnStop.Enabled** blocks to toggle their active status

Don't forget

A component is active when its Enabled property is set to **true** – but inactive when its Enabled property is **false**.

11 Run the application in the emulator or a device then tap the buttons to see them perform their attached behaviors – changing the Label text and toggling the Button states

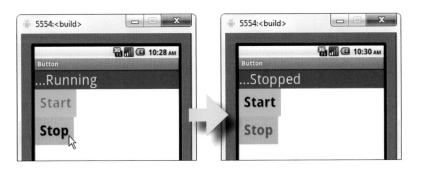

Reading text input

A TextBox component is an essential part of many Android applications, allowing the user to input text for use by the app, where content typed in the TextBox is readable on the screen.

A PasswordTextBox component allows the user to input sensitive text for use by the app, such as a passwords, where content typed in the PasswordTextBox is not readable on the screen.

TextBox.apk

1   Start a new App Inventor project named "TextBox", and also set the screen's Title property to "TextBox"

2   From the Basic Palette, drag a Label, a TextBox, a PasswordTextBox, and a Button onto the Viewer

3   In the Components column, rename the components **lblMessage**, **txtUserName**, **txtPassword** and **btnApply**

4   In the Properties column, edit the components' properties so they resemble the screenshot below:

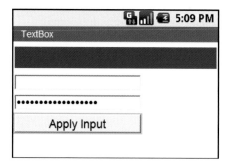

Hot tip

With a TextBox component selected in the Viewer you can add its input hint for the user as the Hint item in the Property column.

5   Now launch the Blocks Editor – ready to add an event-handler to read user input when the Button gets clicked

6   From the My Blocks tab, drag a **btnApply.Click** block onto the workspace

7   Next from the My Blocks tab, snap **set lblMessage.Text**, **set txtUserName.Text** and **set txtPassword.Text** blocks into the Button's Click block on the workspace

8   From the Built-In tab's Text drawer, snap text blocks into
    the **set txtUserName.Text** and **set txtPassword.Text** blocks
    and delete their default content – to assign   empty
    text to each TextBox when the Button gets clicked

9   Next from the Built-In tab's Text drawer, snap a single
    join block into the **set lblMessage.Text** block – providing
    empty sockets for two text blocks to be added

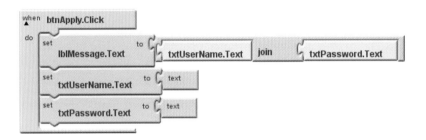

10 Now from the My Blocks tab, snap a **txtUserName.Text**
    block and a **txtPassword.Text** block into the **join** block
    – to assign a concatenated (joined) text string to the
    Label when the Button gets clicked

11 Run the application then enter text in the boxes and click
    the Button to read your text input into the Label

31

Inserting images

An Image component allows images and photos to be displayed in your application. The properties of the Image component specify the actual image file to be displayed and aspects of its appearance, such as width and height.

Supported image file formats are JPG, PNG, GIF, and BMP.

An image is added to an application using the Image component's Picture property. Its Add button can directly specify an image file to be uploaded for inclusion in the app, or select an image file that has been already uploaded as a Media resource.

Image.apk

1. Start a new App Inventor project named "Image", and also set the screen's Title property to "Image"

2. From the Basic Palette, drag two Image components onto the Viewer

3. In the Components column, rename the components **imgRed** and **imgBlue**

4. Select the **imgRed** component, then click its Picture item in the Properties column and click the Add button to launch the Upload File dialog box

5. In the Upload File dialog box, click the Browse button and choose an image file for upload, then click OK to see the image appear in the component on the Viewer

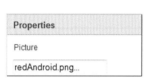

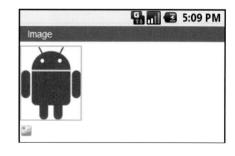

6. In the Media column, click the Add button to launch the Upload File dialog box

7. In the Upload File dialog box, click the Browse button and choose an image file for upload then click OK to see its file name get added to the Media resources list

8. Select the **imgBlue** component, then click its Picture item in the Properties column and choose an image resource from the dropdown list that appears

9. Click OK to see your chosen image appear in the component on the Viewer

Don't forget

The PNG and GIF image file formats both support alpha transparency – so you can have transparent image backgrounds.

10. Run the application in the emulator or in a connected device to see the images appear on the screen

Beware

By default an Image component's Visible property is enabled, but can be disabled by unchecking that item in the Properties column – ensure it is enabled to see the image appear.

Painting canvas

A Canvas component provides a powerful touch-sensitive rectangular area on the screen in which the application can draw shapes and text, and allow user interaction with touch and drag.

Shapes can be drawn on a Canvas by the application when the app first launches using the screen's Initialize event-handler block.

Canvas.apk

1 Start a new App Inventor project named "Canvas", and also set the screen's Title property to "Canvas"

2 From the Basic Palette, drag a Canvas component onto the Viewer

3 In the Components column, rename the component **canBullseye**

4 In the Properties column, set both the Width and Height to 200 pixels, and set the BackgroundColor to Blue – to specify the component's appearance

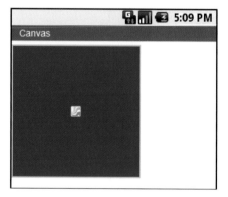

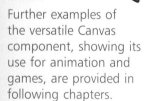

Hot tip

Further examples of the versatile Canvas component, showing its use for animation and games, are provided in following chapters.

5 Now in the Properties column, set the PaintColor to Red – to specify the brush color with which to begin drawing

6 Next launch the Blocks Editor – ready to add an event-handler to draw on the Canvas when the app initializes

7 From the My Blocks tab, drag a **Screen1.Initialize** block onto the workspace from the Screen1 drawer

8 Next drag a **call canBullseye.DrawCircle** block from the canBullseye drawer and snap it into the **Initialize** block

9 Snap **number 100** blocks into each of the three sockets of the **canBullseye.DrawCircle** block - to draw a circle of 100 pixels radius, centered at X:100, Y:100 in the Canvas

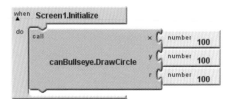

10 Now snap a **set canBullseye.PaintColor** block into the **Initialize** block (below the **DrawCircle** block) and snap a **color Yellow** block into its socket to change brush color

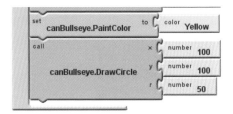

11 Snap a second **call canBullseye.DrawCircle** block into the **Initialize** block (below the **PaintColor** block) and snap number values of 100, 100, and 50 into its sockets - to draw a circle of 50 pixels radius, centered in the Canvas

Hot tip

Click on the workspace and type **Yellow** to create a Yellow color block, or drag one from the Built-In tab's Colors drawer.

12 Run the application in the emulator or a connected device to see the Canvas appear with the specified appearance and see colored circles drawn at the specified coordinates

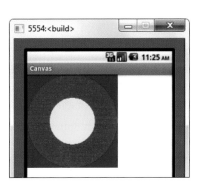

Picking list items

A ListPicker component provides the user an easy way to select an item from a list of text options in an Android application. Tapping a ListPicker in the interface reveals the list of options to the user and an AfterPicking event occurs when an item gets selected. The application may then respond by calling an AfterPicking "event-handler" to perform a task, such as write the selection in a Label component.

ListPicker.apk

 Start a new App Inventor project named "ListPicker", and also set the screen's Title property to "ListPicker"

 From the Basic Palette, drag one Label component and one ListPicker component onto the Viewer

 In the Components column, rename the components **lblMessage**, and **lprColors** respectively

 In the Properties column, edit the components' BackgroundColor, FontSize, and Text properties so they resemble the screenshot below:

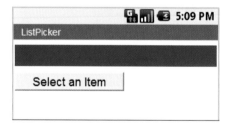

 Select the ListPicker component, then in the Properties column type a comma-separated list of colors into the ElementsFromString field – to specify a list of options

ElementsFromString

Red, Green, Blue

 Next launch the Blocks Editor – ready to add an event-handler behavior for the AfterPicking event

7 From the **lprColors** drawer in the My Blocks tab, drag a **when lprColors.AfterPicking** block onto the workspace

8 Next snap a **set lblMessage.Text** block into the empty socket of the **AfterPicking** block

9 Now snap a **lprColors.Selection** block into the **Text** block – to assign the selected text value to the Label

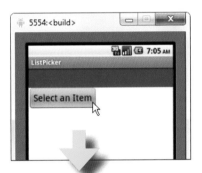

10 Run the application in the emulator or a connected device then choose an option from the available list to see your selection appear in the Label

Beware

Do not confuse the **ListPicker.Selection** block, which contains selected option text value, with the **ListPicker.Text** block, which contains the button face text value.

37

Hot tip

Examples demonstrating how to create option lists in the Blocks Editor are provided in following chapters.

Checking boxes

A CheckBox component allows the user to optionally select individual items in an Android application by checking a box. The Text value of the CheckBox can be processed by the app.

When a CheckBox has been checked its state is set to **true**, otherwise its state is **false**. The action of checking a CheckBox therefore changes its state and so fires a Changed event. The application may then respond by calling a Changed "event-handler" to perform a task, such as write the CheckBox's associated Text value in a Label component.

CheckBox.apk

1. Start a new App Inventor project named "CheckBox", and also set the screen's Title property to "CheckBox"

2. From the Basic Palette, drag one Label component and one CheckBox component onto the Viewer

3. In the Components column, rename the components **lblMessage** and **chkItem**

4. In the Properties column, edit the components' properties so they resemble the screenshot below:

5. Now launch the Blocks Editor – ready to add an event-handler to write the CheckBox's associated Text and its current state into the Label whenever it gets changed

6. From the My Blocks tab, drag a **when chkItem.Changed** block onto the workspace and snap a **set lblMessage.Text** block into its socket

7. Next snap a **chkItem.Text** block into a **join** block (from the Built-In tab) then snap them both into the **Text** block

8 From the Built-In tab, drag another **join** block onto the workspace then snap a **text** block into its first socket and edit is content to read " Checked = "

9 Now from the My Blocks tab, snap a **chkItem.Checked** block into the second join block's empty socket

10 Snap the whole second **join** block into the empty socket on the first **join** block – to complete the event-handler

11 Run the application in the emulator or a connected device then check and uncheck the box to see the app response

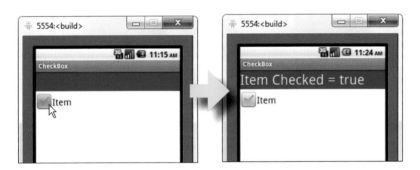

Hot tip

Examples demonstrating how to use only values from CheckBoxes that have been checked are provided in following chapters.

Storing data

A TinyDB component is a non-visible component that provides an easy way to dynamically store and retrieve data in an Android application. The data must be assigned a tag name of your choice when it gets stored, then can be recalled using your chosen name. TinyDB is a persistent data store for the app – so the data stored there will be available each time the app is run. For example, a game's high score can be retrieved each time the game is played.

TinyDB.apk

1 Start a new App Inventor project named "TinyDB", and also set the screen's Title property to "TinyDB"

2 From the Basic Palette, drag one Label, one TextBox, two Buttons, and one TinyDB component onto the Viewer

3 In the Components column, rename the components **lblMessage**, **txtInput**, **btnSave** and **btnRestore** respectively

4 In the Properties column, edit the visible components' properties so they resemble the screenshot below:

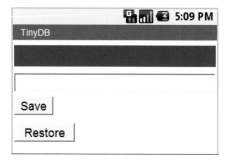

Don't forget

When you drop a TinyDB component onto the Viewer it gets added to the Non-Visible Components list – below the Viewer's screen area.

5 Now launch the Blocks Editor and from the My Blocks tab drag a **when btnSave.Click** block onto the workspace, then snap a **call TinyDB1.StoreValue** block inside it

6 Add a **text** block with a name of your choice to the **StoreValue** block's **tag** socket and a **txtInput.Text** block to its **valueToStore** socket – to assign current text to a tag

7 Add a **set txtInput.Text** block with an empty **text** block – to clear all current text from the TextBox

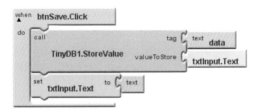

8. Now drag a **when btnRestore.Click** block onto the workspace, then snap a **set lblMessage.Text** block inside it

9. Add a **call TinyDB1.GetValue** block with a **text** block specifying the tag name of the stored data item – to display its content in the Label

10. Run the application then enter some data into the text field, and use the buttons to save and restore its content

41

Telling the time

A Clock component is a non-visible component that fires a Timer event at a regular interval in an Android application. The Timer event is typically set to an interval of 1000 milliseconds (1 second) so could be used to update the seconds counter of a clock display.

Timers can be stopped and started by disabling/enabling their Clock component – to imitate the action of a stopwatch counter.

Clock.apk

1 Start a new App Inventor project named "Clock", and also set the screen's Title property to "Clock"

2 From the Basic Palette, drag one Label component, one Clock component, and two Buttons onto the Viewer

3 In the Components column, rename the components **lblMessage**, **btnStart**, and **btnStop** respectively

4 In the Properties column, edit the visible components' properties so they resemble the screenshot below:

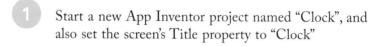

5 Select the Clock component and set its Properties to be enabled and always fire at an interval of 1000 milliseconds

6 Next launch the Blocks Editor and from the My Blocks tab drag a **when Clock1.Timer** block onto the workspace

7 Now snap a **set lblMessage.Text** block into the **Timer** block then add a **+** block from the Built-In tab's Maths drawer

8 Snap a **lblMessage.Text** and Math **number 1** block into the **+** block – to increment the value when the timer fires

9 Drag a **when btnStart.Click** block onto the workspace, then snap a **set lblMessage.Text** block inside it with a **number 0** block – to reset the value to zero

10 Add a **set Clock1.TimerEnabled** block with a **true** Logic block from the Built-In tab – to start the timer

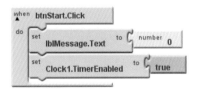

11 Now drag a **when btnStop.Click** block onto the workspace, then snap a **set Clock1.TimerEnabled** block with a **false** Logic block from the Built-In tab – to stop the timer

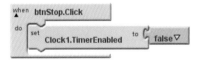

12 Run the application in the emulator or a connected device and use the buttons to control the timer's counter

Configuring screens

Components added to the Viewer are, by default, arranged vertically – one above the other at the left edge of the screen. Interface layout can be better controlled using components from the Screen Arrangement Palette. This provides invisible containers that arrange the components they contain vertically, horizontally, or within cells of a table grid.

Arrangement.apk

1 Start a new App Inventor project named "Arrangement", and also set the screen's Title property to "Arrangement"

2 From the Screen Arrangement Palette, drag a HorizontalArrangement component onto the Viewer

3 Now from the Basic Palette, drag a TextBox and a Button inside the HorizontalArrangement component – to see the container arrange them side-by-side

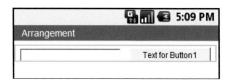

4 Next from the Screen Arrangement Palette, drag a TableArrangement component onto the Viewer – by default this will typically have a 2x2 grid of cells

Don't forget

You can adjust the number of table rows and columns in the TableArrangement component's Properties column.

5 From the Basic Palette, drag Buttons inside each cell of the TableArrangement component – to see the container arrange them in a grid layout

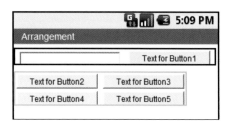

Screen content can also be centered by clever use of the Screen Arrangement components and "padding" by Label components.

6 From the Screen Arrangement Palette, drag a HorizontalArrangement component onto the Viewer and drag a VerticalArrangement component inside it

7 Now from the Basic Palette, drag two Labels into the HorizontalArrangement component – dropping one on each side of the VerticalArrangement component

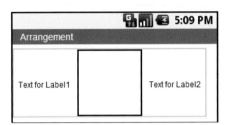

Hot tip

As you drag components onto Screen Arrangement containers a marker appears in the Viewer to indicate how they will be arranged.

8 Select the HorizontalArrangement component then in the Properties column change its Width to "Fill parent..." – to see it expand to fit the available screen width

9 Now with each Label component, delete the default content from their Text property and also set their Width property to "Fill parent..." – to see them expand to center the VerticalArrangement component on the screen

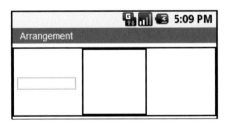

45

Further components placed within this VerticalArrangement component will now appear centered on the screen:

Beware

This centering technique requires that the Width property of the Labels' parent container (i.e. the HorizontalArrangement) must be "Fill parent...".

Summary

- Tapping a Button in the interface creates a Click event to which the app may respond using an event-handler

- A Button's Enabled property can be toggled on or off to make the Button active or inactive as required by the app

- TextBox and PasswordTextBox components both allow the user to input text for use by the app

- An Image component allows images and photos to be used in the app by specifying the name of an image file to display

- Images can be added to the app using the Image component's Property column, or the Add button in the Media column

- A Canvas component provides a rectangular area on the screen in which the app can draw shapes and text

- Shapes can be drawn on a Canvas when the app launches using the screen's Initialize event-handler

- Selecting an option from a ListPicker creates an AfterPicking event to which the app may respond using an event-handler

- Checking a CheckBox creates a Changed event to which the app may respond using an event-handler

- A TinyDB component allows the app to dynamically store and retrieve data using an assigned tag name of your choice

- Data stored in a TinyDB is persistent so will be available each time the app is run

- A Clock component fires an event at a regular interval to which the app may respond using an event-handler

- Timers can be stopped and started by disabling/enabling their Clock component

- Interface layout can be better controlled using components from the Screen Arrangement Palette

3 Controlling progress

This chapter demonstrates the mechanics of programming structures that allow data to be stored, controlled, and manipulated, to progress an Android app.

Composing programs

An application is simply a series of program instructions that tell the device what to do. Although programs can be complex each instruction is generally simple. The device starts at the beginning and works through, instruction by instruction, until it gets to the end. Here are the essential elements of Android app programs:

Statements

A statement is an instruction that performs a program task. For example, the statement **set Screen1.BackgroundColor to color Red** sets the background color of the **Screen1** component to **Red**.

Procedures

A procedure is a group of one or more statements that may be called upon at any time for execution by the program. For example, the statement **call writeLabels** calls upon a procedure to execute statements that might assign values to Label components.

Variables

A variable is a named container defined in the program that stores a value. For example, the statement **def Message text Hello World!** stores a string of text characters in a variable called **Message**.

Operators

An operator is an arithmetical symbol. For example, there are **+** addition, **-** subtraction, **x** multiplication, and **/** division operators.

Objects

An object is a program's fundamental "building block" entity. It can be visible, like a Button, or invisible like a Timer component.

Properties

A property is a characteristic of an object. For example, the property **Label1.Text** is the **Text** property of the **Label1** object.

Methods

A method is an action that an object can perform. For example, the method **Button1.Click** is the **Click** method of the **Button1** object.

Comments

A comment is a note describing the purpose of a statement. For example, "Clear the list" might describe a call to a **Clear** procedure.

Hot tip

The examples in this chapter demonstrate the various elements of a program. Refer back here for identification.

Don't forget

The App inventor IDE is a safe environment in which to experiment and learn from your mistakes.

The illustration below shows the Blocks Editor view of App Inventor programming code for the Click event-handler of a Button component.

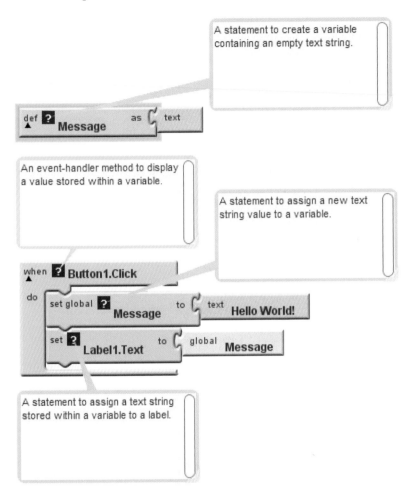

A statement to create a variable containing an empty text string.

An event-handler method to display a value stored within a variable.

A statement to assign a new text string value to a variable.

A statement to assign a text string stored within a variable to a label.

Hot tip

Click the ? spot on the blocks to add comments in the speech bubbles.

Hot tip

You can click on the ▲ black triangle icons that appear on some blocks to collapse or expand parts of the program.

Blocks in App Inventor are colored to indicate particular parts of the program – to differentiate event-handler methods, procedures, variables, numeric values, text values, boolean values (true/false), control routines, and list methods.

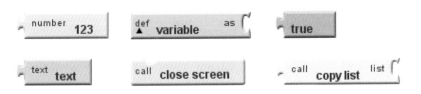

Defining variables

In Android App programming a "variable" is simply a useful container in which a value may be stored for subsequent use. The stored value may be changed (vary) as the program executes its instructions – hence the term "variable".

A variable is created (defined) by first dragging a **def variable** block from the Built-In tab's Definition drawer onto the workspace. The variable must then be given a name by editing the block's default name. The given name should indicate the nature of its contents and, most importantly, the name must be unique within that program – duplicated names are strictly not allowed. Once named, the variable can then be assigned an initial value of any data type – text, number, or boolean (true or false).

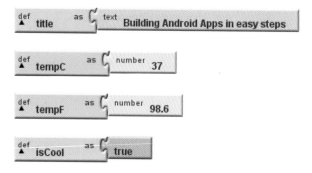

For each defined variable two blocks get automatically added to the My Blocks tab's My Definitions drawer that can be used to get the value stored inside that variable, or to set it a new value.

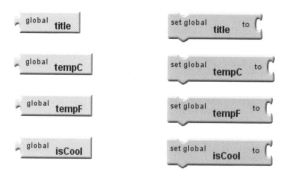

The description on the face of the new blocks contain the term "global" to indicate that the variable is accessible globally within the program – any part of the program can get or set its value.

1 Start a new App Inventor project named "Variable" then add one Label component and one Button component

Variable.apk

2 Launch the Blocks Editor then drag a **def variable** block from the Built-In tab's Definition drawer onto the workspace

3 Edit the default name in the **def variable** block with a name of your choice, then add a text block to set an initial value

4 Next add an event-handler block for the screen's Intialize event that assigns the variable value to the Label

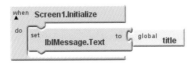

5 Now add an event-handler block for the button's Click event that assigns a new text value to the variable value, then assigns that new variable value to the Label

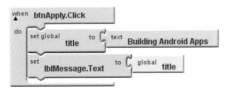

Don't forget

Simply click on the name in a block to set it active so it can be edited.

6 Finally run the app to see the initial variable value appear in the Label upon launch, then click the Button to see the new variable value appear in the Label

Performing operations

Operators in the Built-In tab's Math drawer allow an Android App to progress by performing arithmetical calculations with two given numerical values (operands), or by comparing two operands.

Arithmetical operators

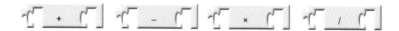

The arithmetical operators **+**, **-**, **x**, and **/** (divide) return the result of an operation performed on two given operands, and act as you would expect. For example, the expression **5 + 2** returns **7**.

Additionally the **modulo** function divides the first operand by the second operand and returns the remainder of the operation. For example the expression **32 modulo 5** returns **2** – five divides into thirty-two six times, with two remainder.

Care must be taken with complex expressions, which contain multiple arithmetical operators, to ensure that operations are performed in the required order to avoid undesirable results. Consider the expression **8 + 4 x 2** for example. Performing operations in left-to-right order **8 + 4 = 12**, then **(12) x 2 = 24**. But performing in right-to-left order **2 x 4 = 8**, then **(8) + 8 = 16**.

Precedence.apk

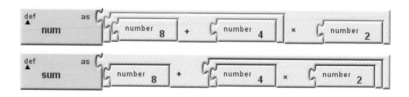

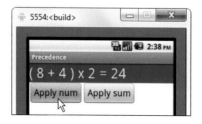

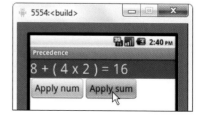

Comparison operators

The comparison operators compare two given operands and return a single boolean value of **true** or **false** – describing the result.

The = equality operator returns **true** when both operands <u>are</u> of equal value, otherwise it will return **false**. Conversely, the **not =** inequality operator returns **true** when both operands <u>are not</u> equal.

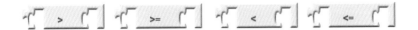

The > "greater-than" operator only returns **true** when the first operand is greater in value than the second operand, whereas the < "less-than" operator only returns **true** when the first operand is less in value than the second operator. Subtly different, the >= and <= operators work in a similar way but also return **true** when both operands are exactly equal.

Operators in the Built-In tab's Logic drawer allow an Android App to progress by performing comparisons between "boolean" operands. These operands are, or can convert to, **true** or **false** values.

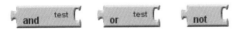

The **and** operator will only return **true** when <u>both</u> operands are themselves **true**, whereas the **or** operator will only return **true** when <u>either one</u> of the operands are themselves **true**.

Usefully, the unary **not** operator returns the inverse boolean state of the supplied operand – reversing **true** to **false**, and **false** to **true**.

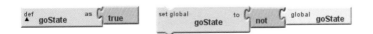

Branching flow

The Built-In Control drawer contains a number of blocks that can be used to progress an Android App by making conditional tests. When a tested condition is **true** given statements get executed, otherwise the program moves on to subsequent statements. Conditional tests can also evaluate complex expressions to test multiple conditions for a **true** value using **and** and **or** Logic blocks.

If.apk

1. Start a new App Inventor project named "If" then add one Label, one TextBox, and one Button component

2. Launch the Blocks Editor then drag in an event-handler for the button's Click event from the My Blocks tab

3. Now add blocks to the event-handler – assigning default text to the Label when the Button gets clicked

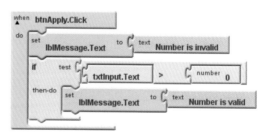

Don't forget

The default "invalid" text will appear in the Label unless the test is found to be true, so the new "valid" text gets inserted.

4. Next drag an **if test** block from the Built-In tab's Control drawer, then add blocks to perform a conditional test – to determine whether the TextBox contains a positive numerical value

5. Now add blocks to the **if test** block – assigning new text to the Label when the test succeeds

6 Run the app and enter a positive number into the TextBox, then click the Button to see the success message

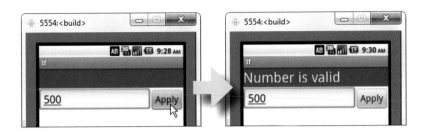

7 Extend the conditional test by inserting an **and test** block from the Logic drawer – to determine whether the TextBox contains a numerical value in the range 1-1000

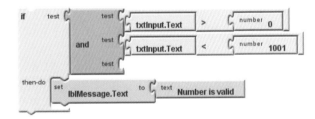

8 Further extend the conditional test by inserting an **or test** block from the Logic drawer – to determine whether the TextBox contains a numerical value in the range 1-1000 or is exactly a value of 2000

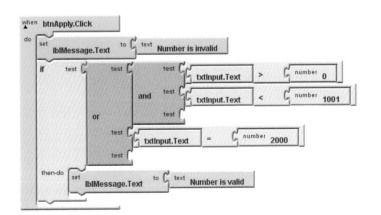

55

Providing alternatives

In addition to the **if test** block, described in the previous example, the Control drawer provides an **ifelse test** block that performs a conditional test, to seek a **true** condition, and also offers an alternative "branch" for the program to pursue when the condition is **false**. In its simplest form this merely nominates an alternative statement to execute when the test fails, but more powerful conditional tests can be constructed by "nesting" **ifelse test** blocks – one inside another. When the program finds a **true** condition it executes those associated statements then immediately exits the nested **ifelse test** blocks, without exploring any further branches.

IfElse.apk

1. Start a new App Inventor project named "IfElse" then add one Label component and one Button component

2. Launch the Blocks Editor then drag a **def variable** block from the Definition drawer on the Built-In tab, and assign it a numeric value of eleven

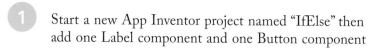

3. Next drag an event-handler for the button's Click event from the My Blocks tab and snap in an **ifelse test** block from the Built-In tab's Control drawer

Don't forget

Conditional branching is the fundamental process by which all computer programs proceed.

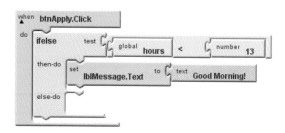

4. Now add blocks to the **ifelse test** block's **test** socket to perform a conditional test – testing if the variable contains a numerical value below thirteen

5. Add blocks to the **ifelse test** block's **then-do** socket – assigning new text to the Label when the test succeeds

6 Run the application then click the Button to see the text message appear as the test succeeds

The Control drawer also provides a **choose test** block that works just like the **ifelse test** block but also returns a final value after its given statements have been executed.

Hot tip

7 Extend the conditional test by inserting a second **ifelse test** block into the **else-do** socket of the first **ifelse test**

8 Now add blocks to the second **ifelse test** block's **test** socket to perform a conditional test – testing if the variable contains a numerical value below eighteen

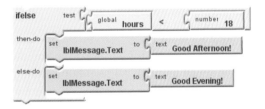

9 Add blocks to the second **ifelse test** block's sockets – assigning new text to the Label when the test succeeds, and assigning new text to the Label when the test fails

10 Adjust the value of the variable upwards, to say fifteen or twenty, then run the app and click the Button to see the appropriate text message appear

Hot tip

The **else-do** statement in the inner **ifelse test** block provides a default statement to execute when both tested conditions are **false**.

57

Notifying messages

The Other Stuff Palette contains a Notifier component that provides pop-up "Alert" dialogs to the user in an Android App. Their drawer in the Blocks Editor contains **ShowAlert** and **ShowMessageDialog** blocks to which text can be simply specified to determine what will be displayed on the dialog box they produce.

Two other dialogs can be provided to gain user input, by choosing from two button alternatives or by entering input into a TextBox.

Dialogs.apk

1 Start a new App Inventor project named "Dialogs" then add a Label, two Buttons, and two Notifier components - naming the Notifier components **dlgChoice** and **dlgInput**

2 Launch the Blocks Editor then, from the My Blocks tab, add a Button click event-handler and from the **dlgChoice** drawer snap in a **call dlgChoice.ShowChooseDialog** block – adding text blocks for the dialog face and its buttons

3 Now add another Button click event-handler and from the **dlgInput** drawer snap in a **call dlgInput.ShowTextDialog** block – adding text blocks for the dialog face

Hot tip

The **ShowAlert** dialog simply displays a message but, unlike the **ShowMessage** dialog, it automatically disappears without requiring the click of an OK button.

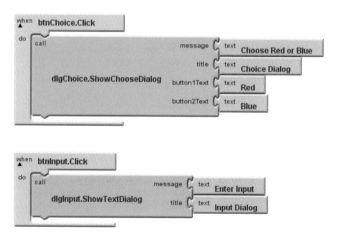

Next add event-handlers to process the user's choice and input:

4 From the **dlgChoice** drawer drag a **dlgChoice.AfterChoosing** block to the workspace – a **name choice** block should be already attached, but can be added manually if absent

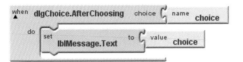

5 Now add a **set lblMessage.Text** block then snap in a **value choice** block from the My Blocks' My Definitions drawer

6 From the **dlgInput** drawer drag a **dlgInput.AfterTextInput** block to the workspace – a **name response** block should be already attached, but can be added manually if absent

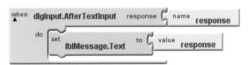

7 Now add a **set lblMessage.Text** block then snap in a **value response** block from the My Definitions drawer

8 Run the app then use the dialogs to get user input

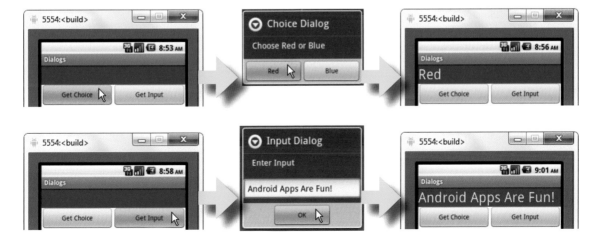

Looping within a range

Programming loops allow an Android App to progress by repeatedly executing specified statements, until the loop ends. Consequently the specified statements must include a test expression to determine when to end – or they will run forever!

The popular **for range** loop uses a counter to test the number of times it has executed (iterated) its statements. This loop ends when it reaches the extreme of its specified range. It is often useful to incorporate the increasing value of the counter into the statements executed on each iteration of the loop.

ForRange.apk

1 Start a new App Inventor project named "ForRange" then add one Label component and one Button component - naming the components **lblMessage** and **btnApply**

2 Launch the Blocks Editor, then add an event-handler for the button's Click event and snap in a **for range** block from the Built-In tab's Control drawer – a loop variable named **i** and a step size of one are created automatically

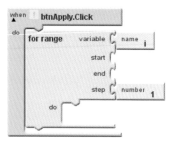

Hot tip

In programming, so-called "trivial" counter variables are typically named **i**, **j**, **k,** and so on.

3 Now add a **number 1** block to the **for range** block's **start** socket and a **number 10** block to its **end** socket – to specify the loop's range, making ten iterations

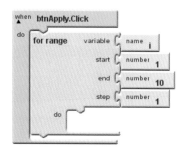

Next add blocks stating what the loop should do on each iteration:

4 Drag a **set lblMessage.Text** block from the My Blocks tab and snap it into the **for range** block's **do** socket

5 From the Built-In tab's Text drawer, add a **call make text** block to the **set lblMessage.Text** block – providing sockets to concatenate (join) values into a single text string

6 Now snap blocks into the **call make text** block to assign a text string to the Label on each iteration, by dragging a **value i** block from the My Blocks' My Definitions drawer and adding a **text** block containing a single space character

Don't forget

Although the counter value is numerical it gets copied into the concatenated string as a text value.

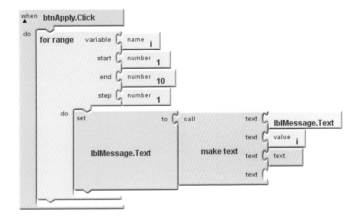

7 Run the app then tap the Button to see the concatenated string get written in the Label – listing the counter value on each iteration of this **for range** loop

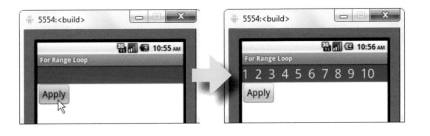

Don't forget

Click on the workspace the type "1" and hit Return to create a **number 1** block. Click on the workspace then type "text" and hit Return to create a **text** block.

Looping through a list

Programming loops in an Android App can also iterate through each item in a List using a **foreach** loop, which executes its specified statements once for each item found in the List. This loop ends when it reaches the end of the List.

ForEach.apk

1 Start a new App Inventor project named "ForEach" then add one Label component and one Button component - naming the components **lblMessage** and **btnApply**

2 Launch the Blocks Editor, then drag a **def variable** block from the Built-In tab's Definition drawer and name the variable "weekdays"

3 Now snap in a **call make a list** block from the Built-In tab's List drawer and add **text** items for each week day

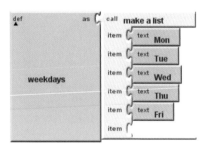

Hot tip

Comprehensive details on List creation and manipulation is provided in chapter Six.

4 Next add an event-handler for the button's Click event to the workspace

5 Snap in a **foreach** block from the Built-In tab's Control drawer – a loop variable named **var** is created automatically

6 Add a **global weekdays** block from the My Blocks' My Definitions drawer to the **in list** socket – to specify the list

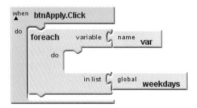

Next add blocks stating what the loop should do on each iteration:

7 Drag a **set lblMessage.Text** block from the My Blocks tab and snap it into the **foreach** block's **do** socket

8 From the Built-In tab's Text drawer, add a **call make text** block to the **set lblMessage.Text** block – providing sockets to concatenate (join) values into a single text string

9 Now snap blocks into the **call make text** block to assign a text string to the Label on each iteration, by dragging a **value var** block from the My Blocks' My Definitions drawer and adding a **text** block containing a space character

Hot tip

A List is also referred to as a variable "array", in which each item is an array "element".

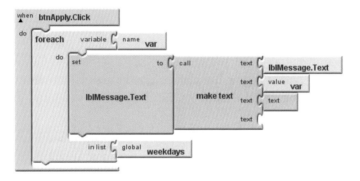

10 Run the app then tap the Button to see the concatenated string get written in the Label – listing the List item value on each iteration of this **foreach** loop

Don't forget

The first string assigned to the **call make text** block is the current value of the Label content.

Looping while true

An alternative to the **for range** loop, a **while** programming loop also allows an Android App to progress by repeatedly executing specified statements, but only while a test condition remains **true**. The **while** loop ends when the test condition becomes **false**.

While.apk

1. Start a new App Inventor project named "While" then add one Label component and one Button component - naming the components **lblMessage** and **btnApply**

2. Launch the Blocks Editor, then drag a **def variable** block from the Built-In tab's Definition drawer – to act as an iteration counter

3. Name the variable **i** and set its initial value to **number 10**

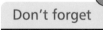

4. Add an event-handler for the button's Click event and snap in a **while** block from the Built-In tab's Control drawer

5. Now add a test expression by inserting a **>=** block from the Built-In tab's Math drawer, a **global i** block from the My Blocks' My Definitions drawer, and a **number 1** block – testing if the counter is greater than zero

Don't forget

The condition could alternatively be tested using the expression **>0**.

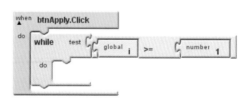

Next add blocks stating what the loop should do on each iteration:

6. Drag a **set lblMessage.Text** block from the My Blocks tab and snap it into the **while** block's **do** socket

7. From the Built-In tab's Text drawer, add a **call make text** block to the **set lblMessage.Text** block – providing sockets to concatenate (join) values into a single text string

8 Now assign a text string to the Label on each iteration, using a **global i** block from the My Blocks' My Definitions drawer and adding a **text** block containing a space

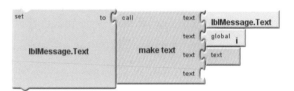

Add a statement to change the counter value on each iteration:

9 From the My Blocks' My Definitions drawer snap a **set global i** block below the **set lblMessage.Text** block

10 Insert a subtraction in the **to** socket using a - block from the Built-In tab's Math drawer, a **global i** block from the My Blocks' My Definitions drawer, and a **number 1** block

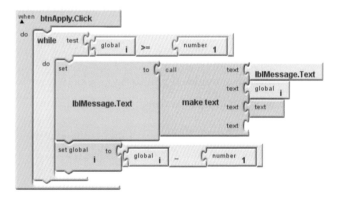

11 Run the app then tap the Button to see the Label display the counter value on each iteration of this **while** loop

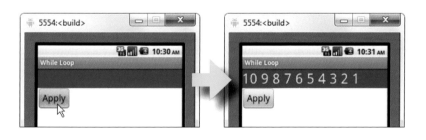

Beware

App Inventor has no "break" or "continue" control blocks – but loops may be exited and iterations skipped by inserting **if** test expressions inside loops.

Hot tip

A **while** loop is often more flexible than a **for range** loop as it tests for a boolean value of **true** or **false**, rather than testing a numeric value.

Summary

- An application is a series of program instructions that tell the device what to do

- A statement is an instruction that performs a program task

- A procedure is a group of one or more statements that may be called upon at any time for execution by the program

- A variable is a container in which a numeric or text string value may be stored for subsequent use by the app

- An object is a fundamental program entity that has Property characteristics and Method actions

- The Math tab provides **+, -, x,** and **/** arithmetic operators that perform a numeric calculation, then return its numerical result

- The Math tab provides comparison operators that perform a numeric comparison, then return its boolean **true** or **false** result

- The Logic tab provides **and, or,** and **not** operators that perform a boolean comparison, then return its boolean **true** or **false** result

- The Control tab provides **if** and **ifelse** blocks that perform conditional branching by testing a given condition

- A Notifier component allows an app to interact with the user via Alert, Message, Choice, and Input dialogs

- The Control tab provides **for range, foreach,** and **while** blocks that repeatedly execute their statements in a loop

- A **for range** loop uses a counter to test the number of times it has executed its statements, and ends when it reaches the extreme of its specified range

- A **foreach** loop executes its statements once for each item in a List, and ends when it reaches the end of that List

- A **while** loop executes its statements while a test condition remains **true**, and ends when that test condition becomes **false**

4 Calling functions

This chapter demonstrates how to group together statements into procedure structures that may be called to perform functions within an Android app .

Calling object methods

Application objects have "methods" that can be called upon to perform particular functions within the app. The method blocks can be found within their component's drawer in the My Blocks tab, and have some differences in the way they are used.

Many methods require information "arguments" to be supplied in the call, such as coordinates for a Canvas object's **DrawCircle** method.

Some methods also return a value to the caller when they are called. For example, a Canvas object's **GetPixelColor** method returns the color of a pixel at the specified location on the Canvas.

Other methods require no arguments and return nothing when called. For example, a Canvas object's **Clear** method simply clears the Canvas area and returns no value to the caller.

Methods.apk

The **text** block shown here contains no text to specify an empty string.

1 Start a new App Inventor project named "Methods" then add a Label, two Buttons, and a Canvas component

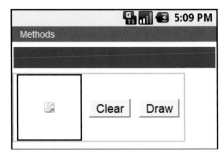

2 Launch the Blocks Editor then drag in an event-handler for the first button's Click event from the My Blocks tab

3 From the Canvas component's drawer snap in a call to its **Clear** method – to clear any Canvas content when called

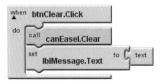

4 Next add a statement to clear any Label content by assigning it an empty text string

5 Now drag in an event-handler for the second button's Click event from the My Blocks tab

6 Snap in a call to the Canvas object's **DrawCircle** method and specify three required arguments – to draw a 50-pixel radius circle when called at coordinates of X:50, Y:50

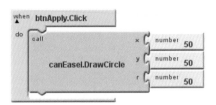

Hot tip

Sockets are provided where arguments are required and App Inventor will warn you of an error if a required argument is omitted.

7 Add a statement with a call to the Canvas object's **GetPixelColor** method – to write the numeric value of the Canvas's central pixel on the Label when called

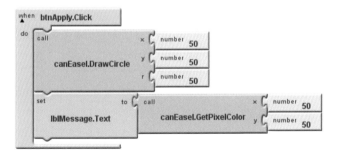

8 Run the app then tap the Buttons to see the Canvas methods draw a circle and write on the Label

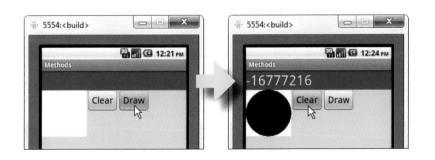

Hot tip

Insert a **call split color** block from the Built-In Colors drawer to translate the somewhat meaningless numeric return from the **GetPixelColor** method into meaningful Red, Green, Blue, and Alpha components 0-255.

Creating procedures

Multiple statements can be usefully grouped within a "procedure" to create a function with a name of your choice. The procedure can then be called upon by name to execute its statements.

As with object methods, described on the previous page, procedures can be created to accept arguments that pass data from the caller, and may also return a value to the caller, or may simply perform a task requiring no arguments and returning no value.

Procedure.apk

1 Start a new App Inventor project named "Procedure" then add three Labels, a TextBox, and a Button component

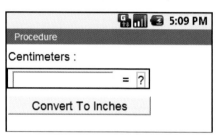

2 Launch the Blocks Editor then drag a **procedure** Definition block onto the workspace and edit its name

3 Next snap an **ifelse** Control block into the procedure block

4 Now **do** add a test using an **is a number?** Math block – to test if the TextBox contains a number

Don't forget

As with variable names, the names given to procedures should be meaningful and unique within that app.

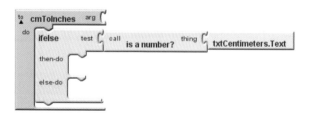

5 Add a statement to the **else-do** socket – to write appropriate advice in the "?" Label when the test fails

6 Next add a statement to the **then-do** socket – to write a computed value in the "?" Label when the test succeeds

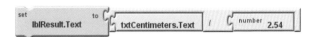

7 Now insert a **format as decimal** Math block in the statement above – to ensure the computed value will always have two decimal places

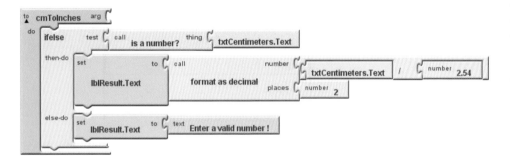

8 Finally drag in an event-handler for the button's Click event from the My Blocks tab, then snap in a **call** block for the named procedure from the My Definitions drawer

9 Run the app then input a value and tap the Button to see the procedure calls respond

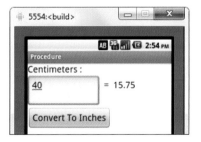

Passing arguments

A powerful feature of procedures is the ability to receive information when they are called. The information is sent as "arguments" in the calling statement, and the receiving procedure must expect the supplied number of arguments. For example, a procedure created to accommodate two arguments must be supplied with precisely two arguments when it is called.

Argument.apk

 Start a new App Inventor project named "Argument" then add two Labels, a TextBox, and a Button component

	🔋📶🔋 5:09 PM
Argument	

Name : [   ]

Apply

 Launch the Blocks Editor then drag a **procedure** Definition block onto the workspace and edit its name

 Next snap a **name** Definition block into the procedure block's **arg** socket and edit its name

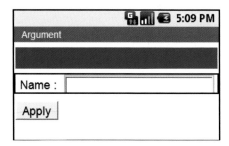

Don't forget

As with variable names, the names given to arguments should be meaningful and unique within the procedure.

 Now **do** add an **ifelse** Control block with a test using an **is text empty?** Text block, and a **value** block from the My Definitions drawer – to test if the TextBox is empty

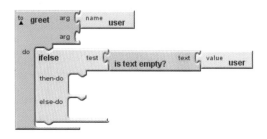

5 Add a statement to the **then-do** socket – to write appropriate advice in the Label when the test succeeds

6 Next add a statement to the **else-do** socket – to write the argument value in the Label when the test fails

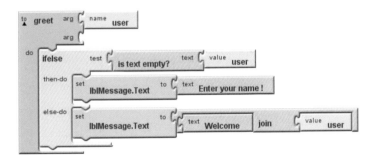

Beware

The **is text empty?** test will only succeed if the TextBox is completely empty – typing letters, numbers, or even space characters will see it fail.

7 Now drag in an event-handler for the button's Click event from the My Blocks tab, then snap in a **call** block for the named procedure from the My Definitions drawer – passing the TextBox contents as its sole argument

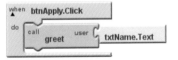

8 Run the app then input a value and tap the Button to see the procedure calls respond using the argument

Hot tip

When you create an argument a **value** block bearing its name gets added to the My Definitions drawer.

Returning results

A procedure can usefully return a value to the caller. This is often desirable to return a result after the procedure has processed values passed to it as arguments. For example, a procedure created to perform arithmetic on two argument values.

Return.apk

1. Start a new App Inventor project named "Return" then add three Labels, two TextBoxes, and a Button component

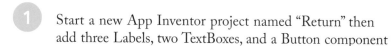

2. Launch the Blocks Editor then drag a variable Definition block onto the workspace and edit its name & value

3. Next drag in a **procedureWithResult** Definition block and edit its name, then snap two **name** Definition blocks into the **arg** sockets and edit their names

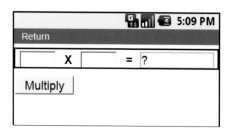

4. Now **do** add an **ifelse** Control block with a test using an **and** Logic block, **is a number?** Math blocks, and **value** blocks from the My Definitions drawer – to test if both arguments are numeric

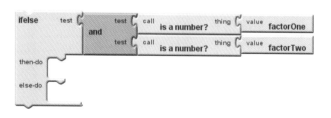

5   Add a statement to the **then-do** socket – to assign a computed value to the variable when the test succeeds

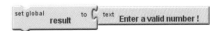

6   Next add a statement to the **else-do** socket – to assign an appropriate message to the variable when the test fails

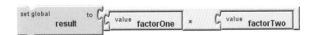

7   Now snap a variable My Definitions block into the procedure's **return** socket – to return its value to the caller

8   Finally drag in an event-handler for the button's Click event from the My Blocks tab, then snap in a **call** block for the named procedure from the My Definitions drawer – passing both TextBox contents as arguments and assigning the returned value to the "?" Label

Hot tip

Notice that no blocks directly relating to the interface appear in the procedure – it is purely manipulating data.

9   Run the app then input values and tap the Button to see the procedure calls respond with the returned value

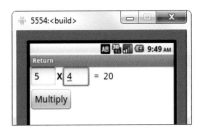

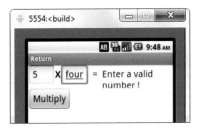

Ignoring results

Occasionally you may wish to call a procedure that returns a value to the caller, yet ignore the returned value. This will typically arise when the procedure is used by more than one app feature. For example, two buttons might call the same procedure but only one uses its returned value. App Inventor has a "|" pipe Definition block that provides a dummy socket for this purpose.

Ignore.apk

1. Start a new App Inventor project named "Ignore" then add a Label, two Buttons, and a Clock component

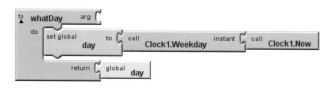

2. Launch the Blocks Editor then drag a variable Definition block onto the workspace and edit its name & value

3. Next drag in a **procedureWithResult** Definition block and edit its name, then **do** add Clock blocks to assign the current weekday number to the variable, and also return that value

The clock's Weekday method returns a number from 1 (Sunday) to 7 (Saturday).

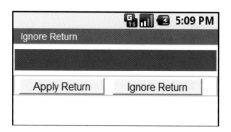

4. Now **do** add an **ifelse** Control block with a test using an **or** Logic block and = Math blocks – to test if the weekday number stored in the variable is either 1 or 7

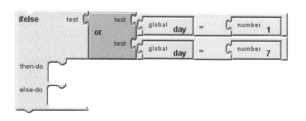

5 Add a statement to the **then-do** socket – to write an appropriate message in the Label when the test succeeds

6 Next add a statement to the **else-do** socket – to write an appropriate message in the Label when the test fails

7 Now add an event-handler for the first button's Click event to write the procedure's return value into the Label

8 Add an event-handler for the second button's Click event

9 Finally **do** add a pipe Definition block and a call to the named procedure from the My Definitions drawer – that will ignore any returned value

10 Run the app then input values and tap the Buttons to see the procedure's return value get used and ignored

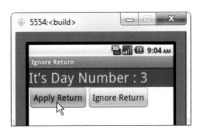 

Beware

The pipe block provides a socket in which to snap a call block - otherwise the call block could not be fitted inside the event-handler block.

77

Don't forget

This procedure writes the Label message but overwrites that where the return value is used.

Calling subroutines

Procedures can be created for the express purpose of providing functionality for other procedures. This is considered to be good practice – as functionality is modularized into separate procedures. A procedure that is called from within another procedure is often referred to as "subroutine". For example, a procedure that returns a text string to its caller might itself call upon a subroutine to perform an arithmetical function.

Subroutine.apk

1 Start a new App Inventor project named "Subroutine" then add a Label, a TextBox, and a Button component

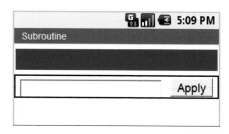

2 Launch the Blocks Editor then drag a variable Definition block onto the workspace and edit its name & value

3 Next drag in a **procedureWithResult** Definition block and edit its name, then add **x** Math blocks and like-named **value** My Definitions blocks – to complete a subroutine that returns its argument cubed

Don't forget

App Inventor insists that the two Math blocks must be nested as shown to determine operator precedence – as described on page 52.

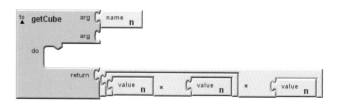

4 Now drag in another **procedureWithResult Definition** block then add an argument **name**, and a variable to return

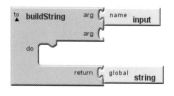

5 Snap an **ifelse** Control block into the second procedure block – to test if the passed argument is numeric

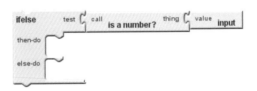

Hot tip

The **join** block is located in the Text drawer.

6 Next add a statement to the **then-do** socket to build a string by employing the subroutine – to write an appropriate message in the Label when the test succeeds

7 Now add a statement to the **else-do** socket – to write an appropriate message in the Label when the test fails

8 Finally add an event-handler for the button's Click event – to pass the current TextBox content to the procedure and to write the procedure's return value on the Label

Hot tip

Consider creating a subroutine if you find yourself repeating similar instructions in an app.

9 Run the app then input values and tap the Button to see the return value include the subroutine's return

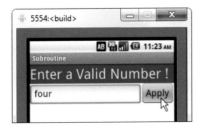
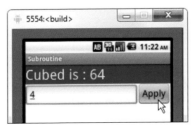

79

Validating input

Procedures can return boolean values to the caller instead of text or numbers. This allows a procedure call to be used directly as a conditional test – where a **true** return will see the test succeed, and a **false** return will see it fail. For example, a procedure may be called upon to validate user input.

Validation.apk

1. Start a new App Inventor project named "Validation" then add a Label, a TextBox, and a Button component

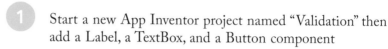
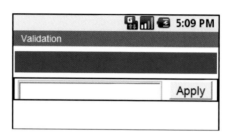

2. Launch the Blocks Editor then drag a variable Definition block onto the workspace and edit its name & value

3. Next add a **procedureWithResult** Definition block that will return the current value of the variable to the caller

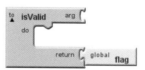

Hot tip

You can find the **contains** block located in the Text drawer.

4. Now **do** snap in an **if** block to test whether the current TextBox value contains an "@" at character

5 Snap another **if** block into the **then-do** socket to test whether the TextBox value contains a "." period character – and to change the variable value when both tests succeed

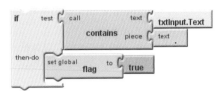

Don't forget

Click on the workspace then type **true** and hit Return to create a boolean block.

6 Next add an event-handler for the button's Click event from the My Blocks tab – to set the variable's initial state

7 Now **do** snap in an **ifelse** Control block – to test if the variable state has changed after validation and write an appropriate message in the Label

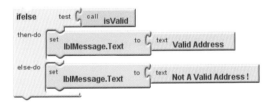

Beware

This example merely checks for the existence of an @ character and a period so does not provide extensive validation of the email format – but it could be developed to become more comprehensive.

8 Run the app then input an email address and tap the Button to see the validation result

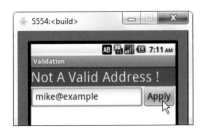
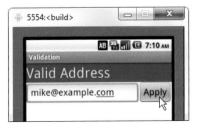

Doing mathematics

The App Inventor Math drawer provides typical mathematic functions that perform as you would expect. Additionally it provides other less familiar functions that can also be useful. For example, to perform circle calculations the constant value of Pi can be assigned to a variable by converting 180 degrees to radians.

APK

Math.apk

1 Start a new App Inventor project named "Math" then add two Labels, a TextBox, and a Button component

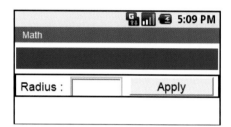

2 Launch the Blocks Editor then drag two variable Definition blocks onto the workspace and edit their name

3 Next add a **convert degrees to radians** Math block to assign Pi, and add an empty **text** block

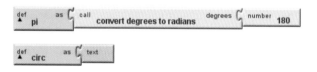

4 Now add an event-handler for the button's Click event

5 Snap in an **ifelse** Control block – to test if the input entered by the user is numeric

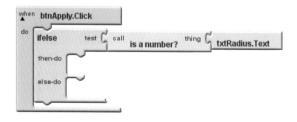

6 Next add a statement to the **then-do** socket to assign a computed value to the empty variable, using Pi

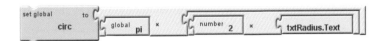

Don't forget

A circle's circumference is calculated by multiplying Pi by its diameter (twice its radius).

7 Now add a statement to the **then-do** socket to ensure the computed value will always have two decimal places

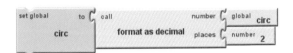

8 Add a statement to the **then-do** socket – to write an appropriate message in the Label when the test succeeds

9 Finally add a statement to the **else-do** socket – to write an appropriate message in the Label when the test fails

10 Run the app then input a number and tap the Button to perform a mathematical calculation

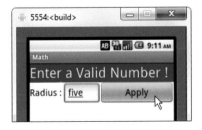
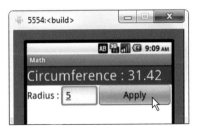

Beware

The value of Pi returned by converting 180 degrees to radians in this example is only correct to five decimal places – as 3.14159.

83

Generating random numbers

Random numbers can be generated using the Math **random integer** function. This returns an integer within a range specified by its two arguments. For example, arguments of one and twenty return a random integer within the range 1-20 inclusive.

Random.apk

1 Start a new App Inventor project named "Random" then add two Labels, a TextBox, and a Button component

2 Launch the Blocks Editor then drag a variable Definition block onto the workspace and edit its name & value

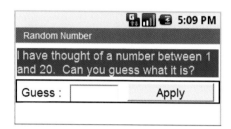

3 Next add a **procedure** to assign a random integer to the variable whenever the procedure gets called

Hot tip

There is also a **random fraction** function that returns a value in the range 0-1 when called.

4 Now call the procedure when the app first launches – to assign an initial random integer to the variable

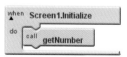

5 Add a Click event-handler containing an **ifelse** Control – to test if an input number is greater than the stored random number, and write an appropriate message when it succeeds

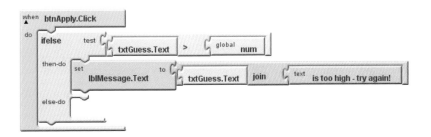

6 Next snap another **ifelse** Control into the **else-do** socket – to test if an input number is less than the stored random number, and write an appropriate message when it succeeds

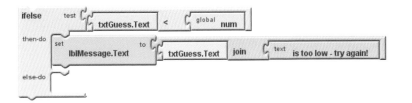

The input test with **is a number?** has been omitted for brevity but should always be included when requesting the user to input a number.

7 Now snap an **if** Control into the remaining **else-do** socket – to test if an input number is equal to the stored random number, and write an appropriate message when it succeeds

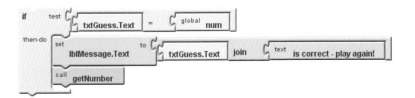

Add a nicety statement to reset the TextBox to empty after each guess – so the user need not do so manually.

8 Add a procedure call to reset the random variable then run the application and guess the random number

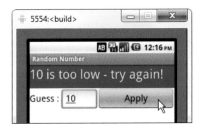
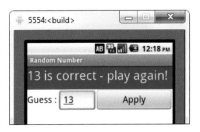

Summary

- Application objects have methods that can be called to perform functions within an app

- Many methods require information to be passed to them as arguments in the call

- Some methods return a value to the caller

- A procedure contains statements to be executed whenever that procedure gets called

- Procedures can be created to accept arguments and return a value to the caller

- Arguments and returns are optional as some procedures merely perform a function

- A procedure requiring arguments must be supplied with the required number of arguments when it is called

- The value returned by a procedure is often the result of processing values supplied as arguments by the caller

- A pipe block allows a procedure to be called and its return value ignored

- Subroutines provide functionality for other procedures

- Procedures can return text values, numeric values, or boolean values to the caller

- A conditional test can directly call a procedure that returns a boolean value when it is called

- The constant value of Pi can be assigned to a variable using a Math function call to convert 180 degrees to radians

- Calling the Math random integer function returns a random number within a range specified by its arguments

5 Managing text

This chapter demonstrates

how to manage text strings

within an Android app.

Manipulating strings

The App Inventor Built-In Text drawer contains the blocks that provide functions to manipulate strings of text. The simple **text** block contains the string "text" by default but this can, of course, be edited to contain any text string the app demands.

A string can be appended to another string by the **join** function. This adds the second given string onto the end of the first given string, concatenating them into a single united string.

Multiple strings can be concatenated into a single string by the **make text** function. This joins all given strings together in order.

Join.apk

1. Start a new App Inventor project named "Join" then add a Label, and a Button component

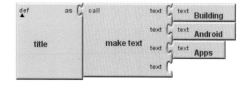

2. Launch the Blocks Editor then drag a variable Definition block onto the workspace and edit its name

Don't forget

When you add to the **make text** block another empty socket appears so the string can be extended.

3. Next snap in a **make text** block – ready to assign string values to the variable

4. Now add simple **text** blocks to build a concatenated string

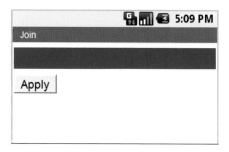

5 Add another variable Definition with a **make text** block then build a second concatenated string

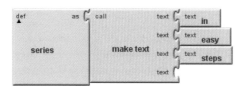

Beware

Remember that each variable here contains a three-word string – the **join** function simply returns a united string.

6 Next create an appending statement with a **join** block – ready to assign a united string to the Label

7 Now drag an event-handler for the button's Click event from the My Blocks tab onto the workspace

8 Snap in the appending statement and add the named variable blocks from the My Definitions drawer

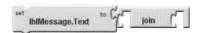

9 Run the app then tap the Button to see the concatenated string get written on the Label

Hot tip

Include a space character after the word in each text block so they will be separated when united.

Querying strings

The Text drawer provides functions that can be used to query strings of text to discover their length and to discover whether they contain a specified piece of text.

A **length** function returns an integer that is the total number of characters within a given string, including space characters.

The **contains** function returns a boolean value of **true** if a specified piece of text is contained in a specified string of text, otherwise it returns **false**.

Also a TextBox component can usefully be tested with an **is text empty?** function that returns **true** if so, otherwise it returns **false**.

Query.apk

1 Start a new App Inventor project named "Query" then add a Label, a TextBox, and a Button component

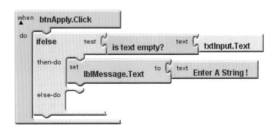

2 Launch the Blocks Editor then drag in an event-handler for the button's Click event from the My Blocks tab

3 Next snap in an **ifelse** Control block to test if the TextBox is empty with an **is text empty?** block

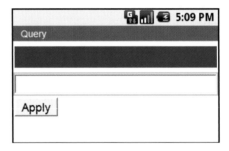

4 Now add a statement in the **then-do** socket – to write a warning message in the Label when the test succeeds

Hot tip

Click on the workspace and type "text" then hit Return to see a default **text** block appear.

5 Snap an **ifelse** Control block into the **else-do** socket to test if the TextBox contains "Java" with a **contains** block

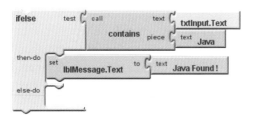

Beware

The search made by the **contains** block is case-sensitive – so that test fails if the TextBox contains "JAVA" or "java".

6 Next add a statement in the **then-do** socket – to write an appropriate message in the Label when this test succeeds

7 Now add a statement in the **else-do** socket – to write an appropriate message in the Label when this test fails

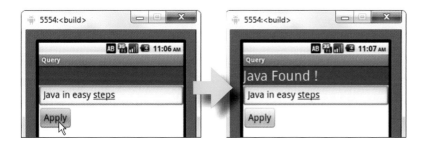

8 Run the app and enter a phrase containing "Java" to see it be identified, then amend the TextBox contents to see the other messages

Don't forget

Remember that the string length total includes any space characters.

Comparing strings

The Text drawer provides functions that can be used to compare text strings to determine their alphabetical order. Interestingly this is achieved by totalling the ASCII code value of each character in the string and noting their order. Uppercase letters A-Z have code values in the range 65-90 whereas lowercase letters a-z have code values in the range 97-122. The total code value and character order will determine the alphabetical order of the strings.

When two identical text strings are compared by the **text=** function their totals are the same so the function returns **true**. The **text<** function only returns **true** when the first string total is <u>less than</u> that of the second string. Conversely the **text>** function only returns **true** when the first string total is <u>greater than</u> that of the second string.

Compare.apk

1 Start a new App Inventor project named "Compare" then add a Label, two TextBox components, and a Button

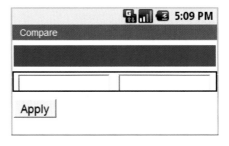

Don't forget

In comparisons character order is taken into account – so comparing "za" to "az" reveals that "za" has a greater total. In terms of ASCII values 'a' is 97 and 'z' is 122.

2 Launch the Blocks Editor then drag in an event-handler for the button's Click event from the My Blocks tab

3 Next snap in an **ifelse** Control block that makes a **text=** comparison between the TextBox contents – and writes an appropriate message on the Label when they match

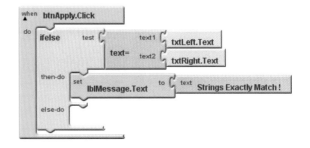

4 Snap an **ifelse** Control block into the **else-do** socket – to make a **text<** comparison between the TextBox contents

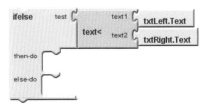

5 Next add a statement in the **then-do** socket – to write an appropriate message in the Label when this test succeeds

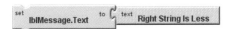

6 Now add a statement in the **else-do** socket – to write an appropriate message in the Label when this test fails

7 Run the app and enter text into both TextBoxes, then tap the Button to compare the strings

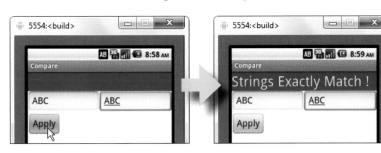

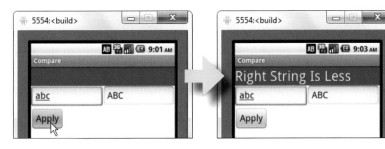

Hot tip

ASCII (pronounced as "as-kee") is the American Standard Code for Information Interchange that provides a standard numerical representation for all characters.

93

Beware

Do not confuse string **text=** and numeric = comparisons – numerically 123 and 0123 are equal, but do not match as strings.

Trimming strings

The Text drawer provides useful funtions to prepare text strings before making a comparison. Any leading or trailing spaces can be removed by the **trim** function. This returns a copy of the string with surrounding spaces removed and is useful to ignore extra spaces that the user may have entered.

A string can be converted to all lowercase by the **downcase** function, or converted to all uppercase by the **upcase** function. These return a converted copy of the string and are useful to allow the user to ignore case when entering text for comparison.

Trim.apk

1. Start a new App Inventor project named "Trim" then add a Label, a TextBox, and a Button component

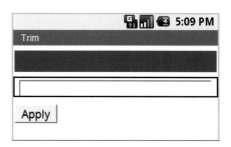

2. Launch the Blocks Editor then drag a variable Definition block onto the workspace and edit its name

Don't forget

You can check that the user has entered something into the TextBox with the **is text empty?** function.

3. Next add an event-handler for the button's Click event that copies the TextBox contents into the variable

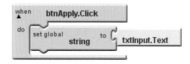

4. Now add a statement within the event-handler to remove leading or trailing spaces from the string in the variable

5 Add another statement within the event-handler to convert the string in the variable to lowercase

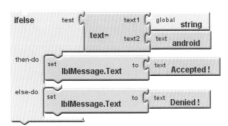

6 Finally within the event-handler add an **ifelse** Control block to compare the string in the variable to "android" – and write an appropriate message in the Label

7 Run the app then, in any keyboard case, enter "Android" with surrounding spaces to see the comparison succeed

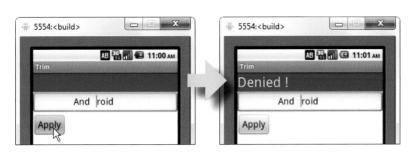

8 Introduce intermediate space characters, or any other text, to see the comparison fail

Splitting strings

The Text drawer provides several functions to split text strings in a variety of ways. Most straightforward of these is **split at spaces** that splits a space-separated text string into individual pieces. These are returned as a list that can be stored in a variable.

An alternative delimiter to a space, such as a comma, can be specified to the **split** function. A comma-separated text string will be returned as a list of pieces, without the comma characters.

A text string can be simply split into two pieces by the **split at first** function, which splits a string around the first occurrence of a specified delimiter. For example, a comma-separated text string will be returned as the first piece and the remainder of the string.

Multiple alternative delimiters can be specified to the **split at any** function and the **split at first of any** function. These both separate a text string around any of their specified delimiters.

Split.apk

1. Start a new App Inventor project named "Split" then add a Label, a TextBox, and a Button component

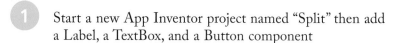

Don't forget

A "delimiter" is a character used to specify a boundary between regions of text.

2. Launch the Blocks Editor then drag a variable Definition block onto the workspace and edit its name

3. Next add an event-handler for the button's Click event – to store a list split from user input within the variable

4 Add a statement to the event-handler to clear the TextBox

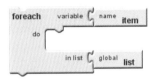

5 Next add a **foreach** Control block to the event-handler – to loop through the list of words stored in the variable

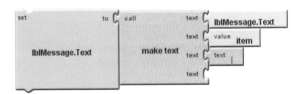

6 Now snap into the **do** socket a statement to write each list piece into the Label, followed by a | pipe character

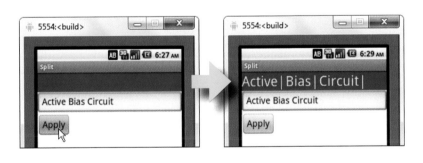

7 Run the app then enter a space-separated string to see the list of pieces get written on the Label

Beware

Remember that the loop variable is <u>named</u> at the start of the **foreach** block, then its <u>value</u> is used inside the loop.

Hot tip

You can write a newline onto a Label by adding a **\n** escape sequence in a text block.

97

Extracting substrings

The Text drawer provides functions that can be used to extract a section from within a string – returning it as a "substring".

A string can be searched for a specified substring using the **starts at** function. This returns an integer that is the substring's first character position in the string, or zero if the search fails. For example, the substring "cut" is found in "Calcutta" at 4.

Specifying the position of a character within a string to the **segment** function, together with a length value, returns a substring that begins at the specified position and is of the specified length.

Multiple occurrences of a substring within a string can all be replaced with a specified alternative by the **replace all** function.

Extract.apk

1. Start a new App Inventor project named "Extract" then add a Label, a TextBox, and a Button component

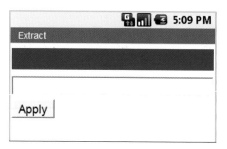

Hot tip

Notice how the **downcase** function is used here to allow a case-insensitive search.

2. Launch the Blocks Editor then drag a variable Definition block onto the workspace and edit its name & value

3. Now add an event-handler for the button's Click event to store the starting position of a substring in the variable

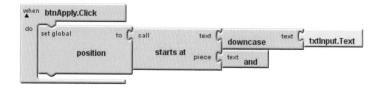

4 Add within the event-handler an **ifelse** Control block to test whether the substring has been located

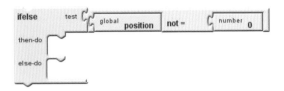

Don't forget

If the variable maintains its initial zero value the **starts at** search has not found the substring.

5 Next add a statement to the **then-do** socket – to write the substring on the Label when the test succeeds

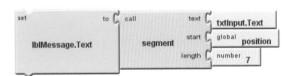

6 Now add a statement to the **else-do** socket – to write an appropriate message on the Label when the test fails

7 Run the app then enter a string and search for the substring to see the extracted substring get written on the Label when it is found within your string

Beware

This search will succeed at the first position it finds the "and" substring – in any string word.

Summary

- The App Inventor Built-In Text drawer contains the blocks that provide functions to manipulate strings of text

- A string can be appended to another string by the **join** function

- Multiple strings can be concatenated into a single string by the **make text** function

- The **length** function returns the size of a string including spaces

- The **contains** function returns **true** if a specified piece of text is found within a string

- A TextBox component can be tested by **is text empty?**

- Comparison of two strings' alphabetical order can be made by the **text<**, **text>**, and **text=** functions

- The **trim** function removes surrounding spaces around a string

- A lowercase copy of a string is returned by the **downcase** function whereas **upcase** returns an uppercase copy of a string

- The **split at spaces** function splits a space-separated string into a list of individual pieces

- The **split** function returns a list of separated string pieces excluding the specified delimiter

- The **split at first** function splits a string into two pieces around the first occurrence of a specified delimiter

- Multiple alternative delimiters can be specified to the **split at any** function and the **split at first of any** function

- The **starts at** function returns the first character position of a substring within a string or zero if the search fails

- The **segment** function returns a substring that begins at a specified string position and is of a specified length

- Multiple occurrences of a substring within a string can all be replaced by the **replace all** function

6 Handling lists

This chapter demonstrates how to create lists and how to handle list items within an Android app.

Making lists

The App Inventor Built-In List drawer contains the blocks that provide functions to create "lists" and to manipulate list items. A list is a series of items that can be assigned to a variable – so the variable can store multiple values, instead of just a single value.

The **make a list** block provides an **item** socket that can accept a **text** block or a **number** block to define the first list item value. Once that is defined the **make a list** block expands to provide further **item** sockets in which to define subsequent list item values.

Assigning a variable that contains a list to a Label component will display all the values as a space-separated list.

List.apk

1. Start a new App Inventor project named "List" then add a Label, and a Button component

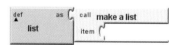

2. Launch the Blocks Editor then drag a variable Definition block onto the workspace and edit its name

3. Next snap in a **make a list** block from the List drawer

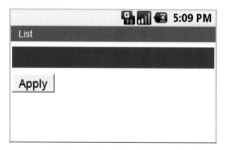

4 Now add a **text** block to define the first list item value – and see the **make a list** block expand

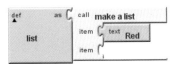

5 Snap in more **text** blocks to define subsequent list items

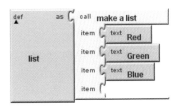

Hot tip

An AppInventor "list" is also known as a variable "array" in other programming languages.

6 Next add an event-handler for the button's Click event

7 Add a statement to the event-handler to display the list

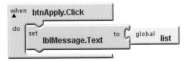

Don't forget

The values appear enclosed within parentheses to identify them as list items.

8 Run the app then tap the Button to see the List items appear as a space-separated list

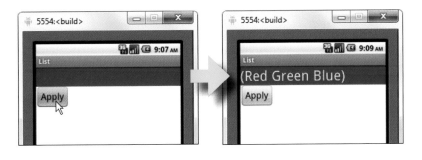

Querying lists

The List drawer provides several functions to query lists in a variety of ways. Most straightforward is the **is a list?** function that returns a boolean **true** or **false** value to determine whether a specified variable does indeed contain a list. Similarly the **is list empty?** function will identify an empty list and the **is in list?** function will seek a specified value, returning a boolean result.

Usefully a list's length can be discovered by the **length of list** function, which returns an integer that is the number of items in the specified list.

Items in a list have indexed position numbers, the first item at position 1. A particular item can be sought in a list by the **position in list** function, which returns an integer that is the index number of that item in the specified list.

QueryList.apk

The creation of this list is described in the example on the previous page.

1. Start a new App Inventor project named "QueryList" then add a Label, and a Button component

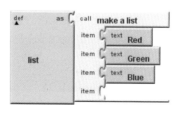

2. Launch the Blocks Editor then drag a variable Definition block onto the workspace and create a list

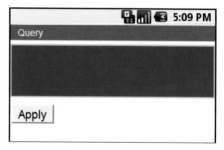

3 Next add an event-handler for the button's Click event
— that tests whether the variable does indeed contain a list

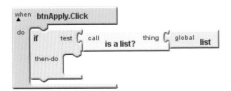

4 Now add a test to this **then-do** socket – to see if a particular item exists

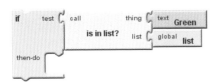

5 Add a statement to this second **then-do** socket to write a two-line message in the Label

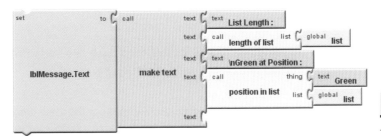

6 Run the app then tap the Button to see the list length and the list index position of the specified item

105

Adding list items

The List drawer provides several functions to add items to a list. An item can be inserted into a list at a particular position by the insert list item function, which specifies the value to be inserted and the index position at which it is to be inserted.

One or more single items can be tacked onto the end of a list starting after the final index position by the **add items to list** function, and a list can be appended to another list using the **append to list** function.

Add.apk

1. Start a new App Inventor project named "Add" then add a Label, and a Button component

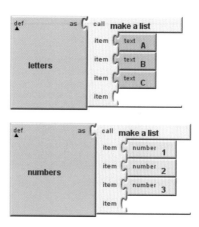

2. Launch the Blocks Editor then drag two variable Definition blocks onto the workspace

Hot tip

The **make a list** function can be used to create an empty list to which items can be added later.

3. Snap into the variable blocks **make a list** List blocks, then add **text** and **number** value blocks to create two lists

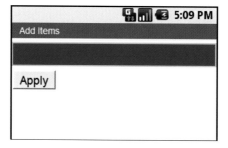

4 Add an event-handler for the button's Click event then snap in an **insert list item** block – to insert a list item at index position 3

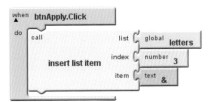

5 Next snap into the event-handler an **add items to list** block – to add a single item at the end of a list

6 Now snap into the event-handler an **append to list** block – to add one list at the end of the other

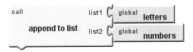

7 Finally, snap into the event-handler a statement to display the modified list then run the app and tap the button to see the result

Selecting items

The List drawer provides a **select list item** function that selects a list item at a specified index position. This can, of course, be used to select a single list item but it can also be used within a loop to select several list items – by incrementing the index number in each iteration of the loop.

Select.apk

1 Start a new App Inventor project named "Select" then add a Label, and a Button component

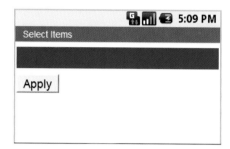

2 Launch the Blocks Editor then drag a variable Definition block onto the workspace and create a list of text items

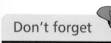

Don't forget

The first item in a list index is at position 1. Other programming languages begin index numbering at zero.

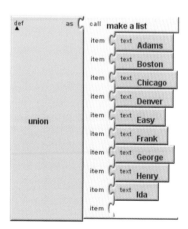

3 Drag a second variable Definition block onto the workspace and create an empty list - to store selected items

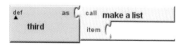

4 Add an event-handler for the button's Click event and snap in a loop – iterating through the text list by steps of 3

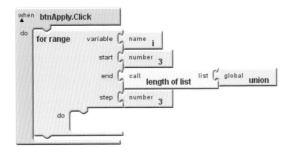

Hot tip

Notice how the **length of list** function is used here to specify when the loop should end.

5 Next snap into the **do** socket a statement to select each third item from the text list and add it to the empty list

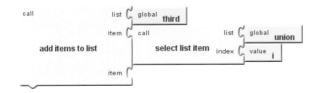

6 Now add a statement to the event-handler to display the items selected by the loop

7 Run the app and tap the Button to see the list of selected items appear on the Label

Swapping items

The List drawer provides a **remove list item** function that removes an item from a list. Subsequent items in the list then adopt new position numbers by shuffling down the index. For example, removing an item at index position 2 means that the item at index position 3 adopts index position 2, and so on.

A list item may also be replaced with a new value using the **replace list item** function, which retains the same index position.

Swap.apk

1 Start a new App Inventor project named "Swap" then add a Label, and a Button component

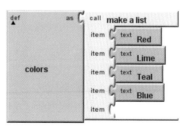

2 Launch the Blocks Editor then drag a variable Definition block onto the workspace and create a list

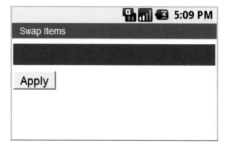

3 Add an event-handler for the screen's Initialize event then snap in a statement to display the initial list items

4 Next add an event-handler for the button's Click event and snap in a statement to remove the third list item

5 Now add a statement to this event-handler to replace the value of the second list item

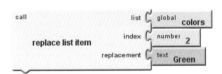

6 Finally add a statement to this event-handler to display the list items once more

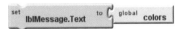

7 Run the app to see the initial list items appear on the Label, then tap the button to see the modified list

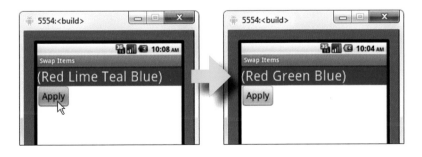

Don't forget

The **Screen1.Initialize** block is located in the **Screen1** drawer under the **My Blocks** tab.

111

Manipulating lists

The List drawer provides a **copy list** function that copies an entire list into another variable. It also provides a **pick random item** function that randomly selects an item from a specified list.

App Inventor usefully has a Built-In Colors drawer that provides blocks of standard named colors that can be used as list items.

Manipulate.apk

1 Start a new App Inventor project named "Manipulate" then add a Label, and a Button component

2 Launch the Blocks Editor then drag two variable Definition blocks onto the workspace and create two lists – one of Color blocks, and one of corresponding names

Don't forget

The actual values of the Color blocks are numeric but have face names for easy recognition.

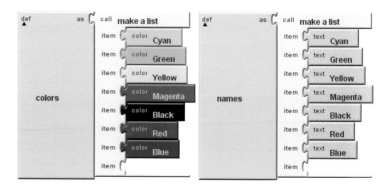

3 Drag another variable Definition block onto the workspace – to store the current selected index position

4 Add an event-handler for the button's Click event to set the Label's background color from a random selection

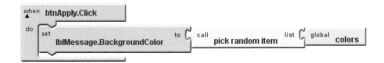

5 Next add a statement to the event-handler to store the index position of the current selected color

6 Now add a statement to the event-handler to set the Label's text name with the current selected color

7 Finally add a test to the event-handler to set the Label's text color – to contrast with the current background color

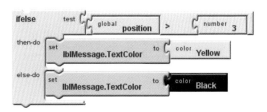

8 Run the app then repeatedly tap the Button to see the Label's background color and text change

Beware

This random selection does not test whether it differs from the current selection – so it may repeat the current selected color.

Separating lists

The List provides functions to separate lists into a string of comma-separated values (CSV). Most straightforward of these is the **list to csv row** function that returns the list items as a comma-separated string with each item surrounded by double quotes. Similarly, the **list to csv table** function returns the list items as a comma-separated string with each item surrounded by double quotes, but also separates each item by a CRLF carriage return. Conversely a list can be created from a comma-separated string assigned to a variable by the **list from csv row** function. Additionally a list can be created from a comma-separated list that also separates each item by a CRLF carriage return (in table format) by the **list from csv table** function.

Separate.apk

1 Start a new App Inventor project named "Separate" then add a Label, and two Button components

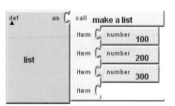

2 Launch the Blocks Editor then drag a variable Definition block onto the workspace and create a list

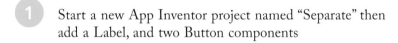

3 Add an event-handler for the screen's Initialize event then snap in a statement to display the initial list items

4 Next add an event-handler for the first button's Click event – to display the list items as a comma-separated row

5 Now add an event-handler for the second button's Click event – to display the list items as a comma-separated table, with each row separated by CRLF carriage returns

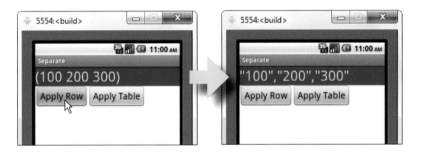

6 Run the app to see the list appear on the Label, then tap the first Button to see the comma-separated list of items

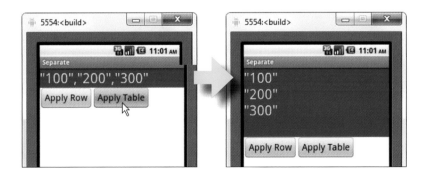

7 Tap the second Button to see the list of items appear on the Label in table format, separated by carriage returns

Summary

- A list is a series of items that can be assigned to a variable so the variable can store multiple values

- The **make a list** block provides **item** sockets that can accept **text** blocks or **number** blocks to define the list item values

- Variables can be queried by the **is a list?** and **is list empty?** functions that return boolean values of **true** or **false**

- The **length of list** and **position in list** functions return an integer value specifying list length and index position

- An item can be inserted into a list at a particular position by the **insert list item** function

- Single items can be added to a list by the **add items to list** function and another list can be added by **append to list**

- The **select a list item** function selects a list item at a specified index position and can be used in a loop to select several items

- The **remove list item** function removes a list item at a specified index position

- A list item can be replaced using the **replace list item** function to specify an index position and a replacement value

- The **copy list** function copies an entire list into another variable

- An item can be randomly selected from a specified list using the **pick random item** function

- The **list to csv row** function return list items as a comma-separated string with each item surrounded by double quotes

- The **list to csv table** function return list items as a comma-separated string with each item surrounded by double quotes and also separated by CRLF carriage returns

- Comma-separated strings can be converted to lists using the **list from csv row** function and **list from csv table** function

7 Embracing media

This chapter describes how to incorporate media resources and animation in an Android app.

Playing sounds

The App Inventor Media components palette provides a Player component that can be used to play audio files. Typically these will be in the popular MP3 file format but Android also supports other audio formats, such as MIDI, WAV, and Vorbis.

A Player component is not visible on the interface but has **Start**, **Pause**, and **Stop** methods, with which to control audio playback, and a **Source** property to specify the audio resource to be played. An audio file must be added to the app as a Media resource, in much the same way that images are added to an app.

Sound.apk

1 Start a new App Inventor project named "Sound" then add three Button components to the interface

2 Next open the Media Palette and drag a non-visible Player component to the interface

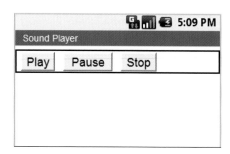

3 Now click the Add button under the Media Components list and select an audio file to upload - see it get added to the Media components list

4 Launch the Blocks Editor then add an event-handler for the screen's Initialize event to specify the audio resource as the Player component's Source – ready for playback

5 Next add event-handlers for the first two button's Click events, to Start and Pause playback of the audio resource

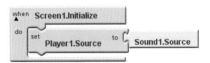

6 Now add an event-handler for the third button's Click event to Stop playback of the audio resource, and to again specify the audio resource as the Player's source – ready for playback once more

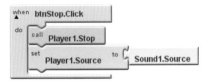

7 Run the app then tap the Buttons to Play, Pause, and Stop playback of the specified audio resource

Playing video

The App Inventor Media components palette provides a Video Player component that can be used to play video files. Android supports the MPEG-4 file format, as well as 3GPP and WMV file formats, but limits the video file size to 1Mb.

A Video Player component appears as a rectangle in the interface. The familiar media controls appear when the user taps the rectangle and are used to call its **Start**, **Pause**, and **SeekTo** methods, to control the video playback. The Video Player component also provides a **Source** property to specify the video resource to play.

A video file must be added to the app as a Media resource, in much the same way that images are added to an app.

Video.apk

1. Start a new App Inventor project named "Video" then add two Button components to the interface

2. Next open the Media Palette and drag a Video Player component to the interface

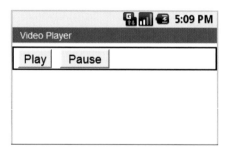

3 Now click the Add button under the Media Components list and select a video file to upload - see it get added to the Media components list

4 With the Video Player component selected, click the empty Source field in its Properties column and choose a video file resource as its source – ready for playback

5 Launch the Blocks Editor then add event-handlers for the two button's Click events, to Start and Pause playback

6 Run the app then tap the Buttons to Play and Pause playback of the specified video resource

Hot tip

Video and Sound source files can be specified to players in the Designer or in the Blocks Editor.

Don't forget

You can also tap the Video block area to reveal the controls to Play and Pause playback.

Snapping photos

The App Inventor Media components palette provides a Camera component that can interact with the phone's camera interface.

A Camera component is not visible on the app interface but has **TakePicture** and **AfterPicture** methods, to snap a photo and to specify an action to perform afterwards.

Photos taken with the **TakePicture** method get added to the phone's gallery, just like any other photo taken by the camera.

Camera.apk

1 Start a new App Inventor project named "Camera" then add a single Button component to the interface

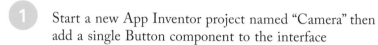

2 Next open the Media Palette and drag a non-visible Camera component to the interface

Hot tip

The Camera component gets added to the Non-visible Components list that appears below the designer Viewer.

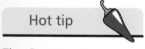

3 Launch the Blocks Editor then add an event-handler for the button's Click event and snap in a call to the **TakePicture** method from the Camera drawer

4 Now add a call to the **AfterPicture** method from the Camera drawer to use the photo as the app's background

5 Click the Package For Phone button in the App Inventor Designer window and download the app to your computer

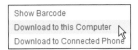

Don't forget

Installation of the app will require you to use the appropriate management software for your particular phone – such as HTC Sync here for HTC phones.

6 Next connect your phone to the computer and install the app on your phone

123

Hot tip

MyPhone Explorer is free phone management software for your computer available from the Android Market and FJ Software at **fjsoft.at**

7 Run the app then tap the Button to snap a photo and set it as the app's background

Picking images

The App Inventor Media components palette provides an Image Picker component that can select an image from the phone's gallery.

An Image Picker component is not visible on the app interface but has an **AfterPicking** method to specify an action to perform after an image has been selected.

The selection of an image assigns the location of that image to the Image Picker component's ImagePath property, and this can be used to specify the image as a background.

BgPicker.apk

1. Start a new App Inventor project named "BgPicker" then add a single Button component to the interface

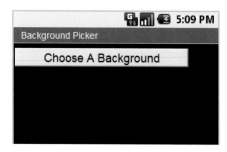

2. Next open the Media palette and drag a non-visible Image Picker component onto the interface

Hot tip

An Image Picker component also has **BeforePicking**, **GotFocus**, and **LostFocus** methods.

3. Launch the Blocks Editor then add an event-handler for the button's Click event and snap in a call to the **AfterPicking** method from the Image Picker drawer

4 Click the Package For Phone button in the App Inventor Designer window and download the app to your computer

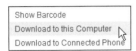

Show Barcode
Download to this Computer
Download to Connected Phone

5 Next connect your phone to the computer and install the app on your phone

6 Now run the app and select an image from your phone's gallery to become the app background

125

Switching screens

At the time of writing App Inventor supports just one Screen component in each app. Multiple "virtual screen" apps can be easily emulated however by controlling the dimensions and visibility of multiple VerticalArrangement components.

Virtual.apk

1 Start a new App Inventor project named "Virtual" then add a VerticalArrangement component and insert a Label and a Button component within it

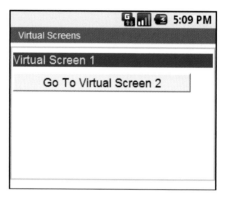

2 Select the VerticalArrangement component in the Viewer then name it "VirtualScreen1" and set its dimension Properties to fill the width and 200 pixels high

3 Next add a second VerticalArrangement component containing a Label and a Button

4 Name this VerticalArrangement component "VirtualScreen2" and set its dimension Properties to fill the width and 200 pixels high – just like the first one

5 Select the second VerticalArrangement component in the Viewer then uncheck its Visible property, to hide it

Properties

Visible

☐

6 Launch the Blocks Editor then add an event-handler for the first screen button's Click event and snap in statements to toggle the visibility of both virtual screens

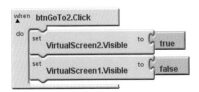

7 Now add an event-handler for the second screen button's Click event and snap in statements to toggle the visibility of both virtual screens

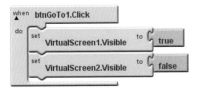

8 Run the app then tap the buttons to switch between the virtual screens

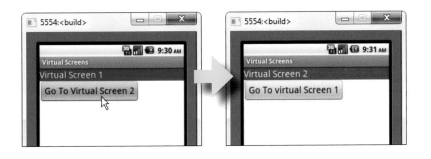

Animating components

The App Inventor Animation components palette provides a Ball component and an ImageSprite component that can be animated. The Ball is a sprite component, which is a filled disc that looks like a ball, but the ImageSprite component displays a specified image. Both these components must be contained within a Canvas component and can react to other sprites and to the Canvas edges.

A sprite's movement is specified by its properties. For example, to have a ball move 5 pixels toward the top of a canvas every 500 milliseconds (half second), you can set the Speed property to 5, the Interval property to 500, the Heading property to 90 (degrees), and the Enabled property to True.

When a sprite hits the edge of the Canvas an EdgeReached event occurs that numerically recognizes which edge has been reached :

Edge:	Value:
North	1
NorthEast	2
East	3
SouthEast	4
South	-1
SouthWest	-2
West	-3
NorthWest	-4

The edge value can then be passed to the sprite's Bounce method to reverse its direction of movement.

Animation.apk

1. Start a new App Inventor project named "Animation" then add a Canvas and a Ball component to the interface

2. Select the Ball component and set its physical properties to PaintColor Red, Radius 20, and check Visible

3. Next set the Ball's movement properties to Speed 15, Interval 100, Heading 25, and check Enabled

4 Now set the Ball's initial coordinate properties to X 25, Y 25, and Z 1.0

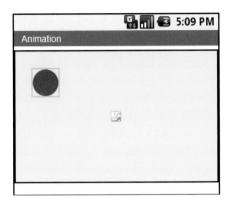

5 Launch the Blocks Editor then add an event-handler for the ball's EdgeReached event – an edge argument is automatically created

6 Finally snap in a call to the ball's Bounce method and specify the value of the edge as its argument

7 Run the app to see the Ball bounce around the Canvas

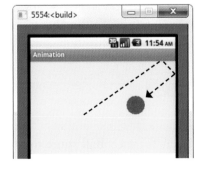

129

Detecting collisions

An animated sprite can be made to react to another animated sprite when their movements collide on the Canvas. At the moment of collision a CollidedWith event occurs. The sprite's CollidedWith event-handler specifies the other sprite as its argument and can be used to change direction of both sprites.

The current direction of sprites is stored in their Heading property as a numeric value of degrees in the range 0-360. Zero is horizontally to the right (East), 90 is straight up (North), 180 is to the left (West), and 270 is straight down (South). These values can be changed in response to a collision to specify a new direction in which to move. Simply adding a numeric value to the current Heading will specify a new direction. In order to ensure the new value does not exceed 360 it is necessary to use the modulo operator. For example, 270+120=390, modulo 360=30.

Collide.apk

1. Start a new App Inventor project named "Collide" then add a Canvas and two Ball components to the interface, and set their properties as those in the previous example

Collide   5:09 PM

Don't forget

The blue Ball sprite must be repositioned so it has a different starting point to that of the red Ball.

2. Launch the Blocks Editor then add an event-handler for the screen's Initialize event to reposition the blue Ball sprite

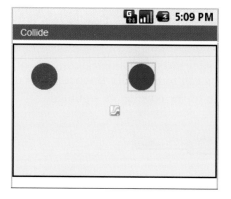

3. Next add event-handlers for each Ball sprite to reverse their direction when they collide with a Canvas edge

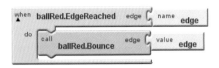

Don't forget

The Bounce method was described in the previous example and automatically creates edge arguments.

4 Now add an event-handler for the blue Ball sprite's CollidedWith event – an argument is automatically created

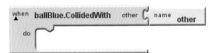

5 Finally snap statements into the CollidedWith event-handler to change the direction of each Ball sprite

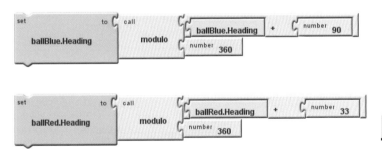

6 Run the app to see the Balls bounce around the Canvas and change direction when they collide with each other

Hot tip

A procedure can be created to change the direction of multiple Ball sprites by passing in the current heading and returning a new value.

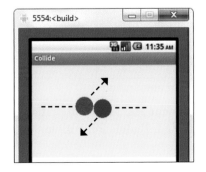

Dragging objects

When the user taps a sprite a Touched event occurs and the sprite's Touched event-handler method can be called to perform an action. For example, to change a Ball sprite's PaintColor or to change an ImageSprite's Picture. The Touched method returns the XY coordinates at which the user has tapped the Canvas.

When the user drags a sprite a Dragged event occurs and the sprite's Dragged event-handler method can be called to reposition the sprite accordingly. The Dragged method returns three sets of XY coordinates:

- **Starting Position** – the location on the Canvas where the user began this dragging action

- **Previous Position** – the location on the Canvas immediately prior to the current position

- **Current Position** – the location on the Canvas where the user is currently dragging the sprite

Assigning the XY coordinates of the current position to the sprite's MoveTo method enables it to move around the Canvas along with the user's dragging action.

Drag.apk

1 Start a new App Inventor project named "Drag" then add a Label, a Canvas, and a Ball sprite to the interface

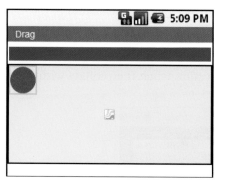

2 Select the Ball component and set its properties to Enabled, PaintColor Red, Radius 20, Visible, X 0, Y 0

3 Launch the Blocks Editor then add an event-handler for the sprite's Dragged event – arguments are automatically created

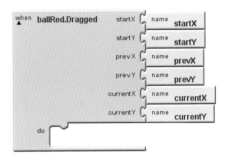

4 Next snap in a statement to move the sprite to the current drag position

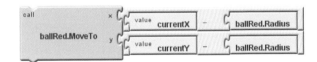

5 Now snap in a statement to display the current location

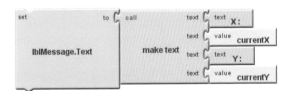

6 Run the app then drag the sprite to see it follow the current XY coordinates displayed on the Label

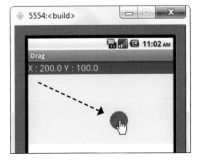

Dropping objects

Dragged sprites can be made to disappear when they collide with another sprite to resemble the action of dropping an object into a recycle bin. Upon collision with another sprite the CollidedWith event fires and can be used to set the dragged sprite's Visible property to false, causing that sprite to vanish.

Where multiple sprites appear on the same Canvas their Z-index value determines which will appear uppermost when they overlap.

Drop.apk

1 Start a new App Inventor project named "Drop" then add a Canvas, a Ball sprite, an ImageSprite, and a Button component to the interface

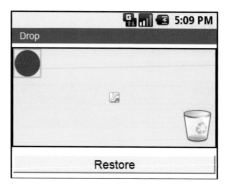

2 Set the Ball sprite properties as in the previous example – so it becomes draggable

3 Next use the Add button in the Media column to specify an image resource for the ImageSprite

4 Now set the Ball sprite's Z property to 2 and the ImageSprite's Z property to 1 – to keep the Ball on top

5 Launch the Blocks Editor then add an event-handler for the Ball sprite's Dragged event as in the previous example

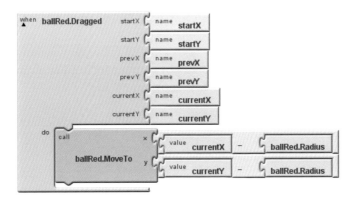

6 Next add event-handlers for the Ball's CollidedWith event and for the button's Click event

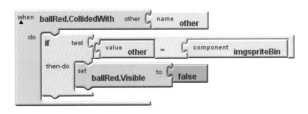

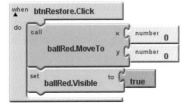

7 Run the app then drop the Ball on the ImageSprite to see it vanish, and click the Button to see it reappear

Beware

The CollidedWith event-handler tests that the collision occurs with the ImageSprite by comparing the <u>value</u> of the other sprite against its <u>component</u> identity.

135

Hot tip

The Ball sprite's Enabled property could also be toggled from true to false to disable it when invisible.

Summary

- The Player component on the Media palette can be used to playback an app's audio files

- The Video Player component on the Media palette can be used to playback an app's video files

- Audio and video files are added to an app as a Media resource in much the same way that images are added to an app

- The Camera component on the Media palette can interact with the phone's camera interface

- Photos taken with the Camera component's **TakePicture** method get added to the phone's gallery like other photos

- The ImagePicker component on the Media palette can select an image from the phone's gallery

- Multiple "virtual screens" can be emulated by controlling the visibility of multiple Vertical Arrangement components

- A Ball sprite is a filled disc that resembles a ball whereas an ImageSprite component displays a specified image

- Sprites must be contained within a Canvas component and can be animated by specifying their Speed, Interval, and Heading

- When a sprite hits the edge of the Canvas an **EdgeReached** event occurs that numerically recognizes that edge

- The edge value returned by the **EdgeReached** method can be passed to the **Bounce** method to reverse the sprite's heading

- The **CollidedWith** event returns the identity of the other sprite with which it has collided

- When the user taps a sprite a **Touched** event occurs

- Dragging a sprite returns its XY coordinates which can be assigned to its **MoveTo** method so it follows the user's action

- Setting a sprite's **Visible** property to **false** causes it to disappear

8 Sensing conditions

This chapter demonstrates how an app can interact with the phone's interfaces to sense its condition and utilize its social data.

Pin-pointing location

The App Inventor Sensors palette provides a LocationSensor component that can detect the phone's location using a combination of Wi-Fi, carrier network, and GPS techniques.

A LocationSensor component is not visible on the app interface but has a **LocationChanged** method to return the current latitude, longitude, and (if supported by the phone) altitude. It also has a **CurrentAddress** property, to extract the current street location from the coordinates, and a **ProviderName** property to specify which detection technique to use – such as GPS.

Geolocation.apk

 Start a new App Inventor project named "Geolocation" then add a LocationSensor and four Label components

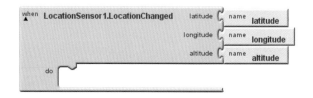

 Launch the Blocks Editor then add an event-handler for the screen's Initialize event and snap in a statement to specify GPS as the detection technique

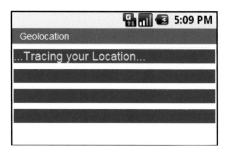

Next add an event-handler for the sensor's LocationChanged event – arguments are automatically created

Hot tip

The LocationSensor component gets added to the Non-visible components list that appears below the designer Viewer.

4 Now add statements to the LocationChanged event-handler to assign text to each of the Label components

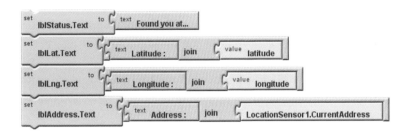

5 Click the Package For Phone button in the App Inventor Designer window and download the app to your computer

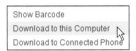

Show Barcode
Download to this Computer
Download to Connected Phone

6 Next connect your phone to the computer and install the app on your phone

Beware

App Inventor cannot presently emulate LocationSensor apps – they must be installed on a phone to run.

7 Run the app to discover the phone's current location

Recognizing orientation

The App Inventor Sensors palette provides an OrientationSensor component that can detect the phone's current orientation.

An OrientationSensor component is not visible on the app interface but has an **OrientationChanged** method that returns three values expressed in degrees:

- **Azimuth** – the phone's relationship to magnetic North, where 0° is North, 90° is East, 180° is South, 270° is West, etc.

- **Pitch** – the phone's angle top-to-bottom, where 0° is level, increasing to 90° degrees as the phone's top is pointed down, and decreasing to -90° as its bottom is pointed down

- **Roll** – the phone's angle left-to-right, where 0° is level, increasing to 90° degrees as the phone's left side is pointed down, and decreasing to -90° as its right side is pointed down

The values returned by the OrientationSensor have several decimal places but it is often useful to round them to whole numbers.

 Start a new App Inventor project named "Orientation" then add an OrientationSensor and three Label components

Orientation.apk

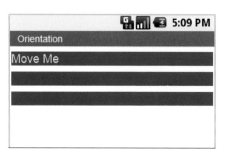

Launch the Blocks Editor then add an event-handler for the screen's Initialize event and snap in a statement to enable the OrientationSensor

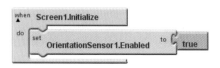

3 Add an event-handler for the sensor's OrientationChanged event – arguments are automatically created

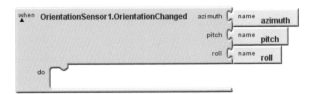

Hot tip

The OrientationSensor also has Angle and Magnitude properties that contain values describing how a ball might roll on the screen.

4 Now add statements to the OrientationChanged event-handler to assign text to each Label component

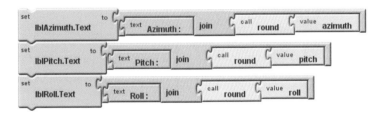

5 Click the Package For Phone button in the App Inventor Designer window and download the app to your computer

Show Barcode
Download to this Computer
Download to Connected Phone

6 Connect your phone to the computer and install the app on your phone

Beware

App Inventor cannot presently emulate OrientationSensor apps – they must be installed on a phone to run.

7 Run the app then move the phone to detect its orientation

141

Feeling movement

The App Inventor Sensors palette provides an AccelerometerSensor component that can detect the phone's movement.

An AccelerometerSensor component is not visible on the app interface but has a **Shaking** method that is called when the phone gets moved vigorously in a shaking motion. This can be used to change an image source in response to the movement.

Acceleration.apk

 Start a new App Inventor project named "Acceleration" then add an AccelerometerSensor and an Image component

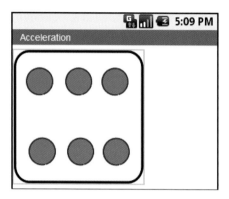

Next use the Add button in the Media column to specify image resources for use by the Image component

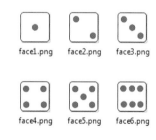

The AccelerometerSensor component gets added to the Non-visible components list that appears below the designer Viewer.

Now launch the Blocks Editor then add an event-handler for the screen's Initialize event and snap in a statement to enable the AccelerometerSensor

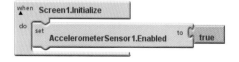

4 Define a variable to store an integer value then add an event-handler for the AccelerometerSensor's Shaking event

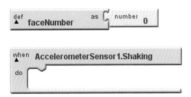

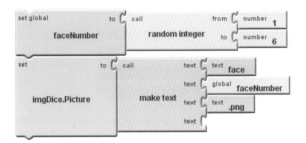

5 Snap in statements to assign a random value to the variable and assign a source to the Image component

143

6 Use the Package For Phone button to download the app to your computer then install the app on your phone

Show Barcode
Download to this Computer
Download to Connected Phone

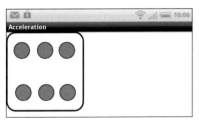

7 Run the app then shake the phone to see the Image change

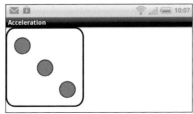

Picking contacts

The App Inventor Social palette provides a number of components to interact with members of your phone's contact list.

An EmailPicker component is simply a TextBox component that automatically completes email addresses from your contact list. More interestingly, the ContactPicker component provides a Button on the interface that launches the phone's contact list. Picking a contact fires the ContactPicker's **AfterPicking** event and stores that contact's relevant details in its **ContactName** and **EmailAddress** properties.

ContactPicker.apk

1 Start a new App Inventor project named "ContactPicker" then add two Labels and a ContactPicker component

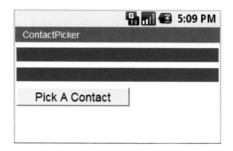

144

2 Next launch the Blocks Editor then add an event-handler for the screen's Initialize event and snap in a statement to enable the ContactPicker

Don't forget

The ContactPicker can alternatively be enabled by checking its Enabled checkbox in the Designer Properties column.

3 Now add an event-handler for the ContactPicker's **AfterPicking** event

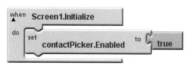

④ Snap in statements to assign the chosen contact's details to the Label components

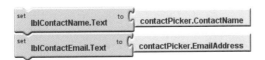

⑤ Click the Package For Phone button in the App Inventor Designer window and download the app to your computer

⑥ Connect your phone to the computer and install the app on your phone

⑦ Run the app then tap the ContactPicker button to open your phone's contact list and pick a contact to see their details appear on the Labels

Hot tip

A ContactPicker also has a Picture property that stores the name of the contact's image file, which could be assigned to an Image component.

Calling phone numbers

The App Inventor Social palette provides a PhoneNumberPicker component that provides a Button on the interface to launch the phone's contact list – much like the ContactPicker described in the previous example. Picking a contact fires the PhoneNumberPicker's **AfterPicking** event and stores that contact's relevant details in its **ContactName** and **PhoneNumber** properties.

The Social palette also has a non-visible Phonecall component that can actually call a number. Its **MakePhoneCall** method dials the number assigned to its **PhoneNumber** property – which can be specified manually or by the PhoneNumberPicker.

PhoneCall.apk

1 Start a new App Inventor project named "PhoneCall" then add two Labels, a PhoneNumberPicker, a Button, and a PhoneCall component to the interface

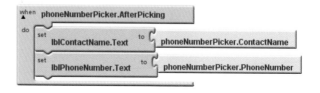

Don't forget

The PhoneCall component gets added to the Non-visible components list that appears below the designer Viewer.

2 Next launch the Blocks Editor then add an event-handler for the screen's Initialize event and snap in a statement to enable the PhoneNumberPicker

3 Now add an event-handler for the PhoneNumberPicker's **AfterPicking** event to display the chosen contact's details

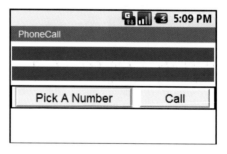

4 Add an event-handler for the button's Click event to dial the phone number of the chosen contact

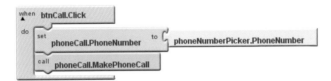

5 Click the Package For Phone button in the App Inventor Designer window and download the app to your computer

6 Connect your phone to the computer and install the app on your phone

7 Run the app then tap the PhoneNumberPicker button to open your phone's contact list, then pick a contact and press the Button to call the displayed contact's number

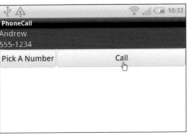

Hot tip

A PhoneNumberPicker also has EmailAddress and Picture properties that store other details of the chosen contact.

Texting messages

The App Inventor Social palette provides a non-visible Texting component that can listen for incoming SMS text messages when the app is running by setting its **ReceivingEnabled** property to **true**. Its **WhenReceived** event-handler then gets passed the phone number and message of incoming text messages.

The Texting component can also send SMS text messages from an app using its **SendMessage** method. The Texting component stores details in **PhoneNumber** and **Message** properties. Its **PhoneNumber** property can be specified manually or by a PhoneNumberPicker.

Texting.apk

1 Start a new App Inventor project named "Texting" then add a TextBox, a PhoneNumberPicker, a Button, and a Texting component to the interface

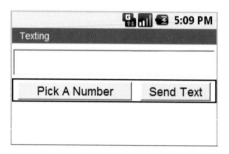

2 Next launch the Blocks Editor then add an event-handler for the screen's Initialize event and snap in a statement to enable the PhoneNumberPicker

3 Now add an event-handler for the PhoneNumberPicker's **AfterPicking** event to store the Texting details

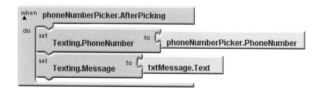

4 Add an event-handler for the button's Click event to send the text to the phone number of the chosen contact

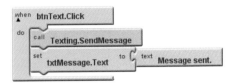

5 Click the Package For Phone button in the App Inventor Designer window and download the app to your computer

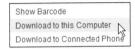

6 Connect your phone to the computer and install the app on your phone

7 Run the app then enter a message, tap the PhoneNumberPicker button to pick a contact, and press the Button to send that contact your message

Hot tip

The PhoneNumber property can include hyphens, periods, and parentheses, but not empty spaces.

Tweeting updates

The App Inventor Social palette provides a non-visible Twitter component that can interact with the Twitter API to both send and receive status messages. In order to use this component the app must be given a unique name and registered for Open Authorization (OAuth) at **http://twitter.com/oauth_clients/new**. Registration provides information required by the Twitter component's **ConsumerKey** and **ConsumerSecret** properties.

Once the **ConsumerKey** and **ConsumerSecret** properties have been set to those provided by OAuth the Twitter component's **Authorize** method can be called to open your account's login page. After logging-in with your user name and password an **Authorized** event fires and the screen returns to the app interface.

The Twitter component can listen for the **Authorized** event with its **IsAuthorized** event-handler and respond when it fires. After authorization the Twitter component's **SetStatus** method can then specify a Text value to submit to your Twitter status.

Tweet.apk

1 Start a new App Inventor project named "Tweet" then add a TextBox and a Button component to the interface

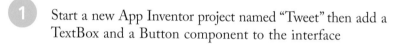

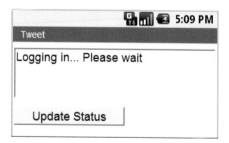

2 Launch the Blocks Editor then add an event-handler for the screen's Initialize event and snap in statements to disable the TextBox, set account values, and Authorize

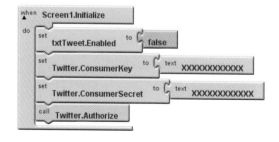

3 Next add an event-handler for Twitter's Authorized event to enable the Textbox and change its displayed message

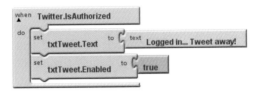

Don't forget

The Twitter component gets added to the Non-visible components list that appears below the designer Viewer.

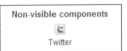

4 Now add an event-handler for the button's Click event to update your Twitter status with a new message

5 Run the app then wait briefly until the login succeeds

6 Finally type a message in the TextBox the tap the Button to see your Twitter status get updated

Hot tip

The Twitter component also has methods and properties to Search and Follow Twitter accounts.

Storing data online

The Other Stuff palette's TinyWebDB component is a non-visible component that can dynamically store and retrieve data in an Android application. Unlike the TinyDB component, which stores data locally on the phone, the TinyWebDB component communicates with a Web service to store the data online – so the stored data can be accessed by multiple permitted apps and devices. The data must be assigned a tag name of your choice when it gets submitted by its **StoreValue** method. The data can subsequently be retrieved using that chosen name with its **GetValue** method.

TinyWebDB.apk

1. Start a new App Inventor project named "TinyWebDB" then add a Label, a TextBox, and two Button components

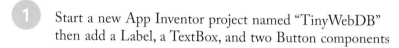

2. Launch the Blocks Editor then add an event-handler for the submit button's Click event and snap in statements to store the TextBox content using a given tag name

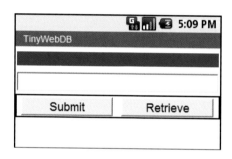

Hot tip

Choose a comprehensive tag name that is explicitly relevant so it is not easily duplicated.

3. Now snap in statements to display the stored TextBox value then remove the TextBox content

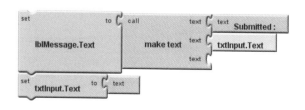

4 Next add an event-handler for the retrieve button's Click event and snap in statements to get the stored value using the given tag name

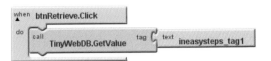

Don't forget

The **value** block for **valueFromWebDB** gets added to the My Definitions drawer when **GotValue** is dropped onto the Blocks Editor.

5 Add an event listener to assign the retrieved tag value – arguments are automatically created for its name and value

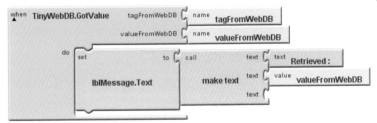

153

6 Run the application then enter some data into the TextBox and use the buttons to submit and retrieve its content

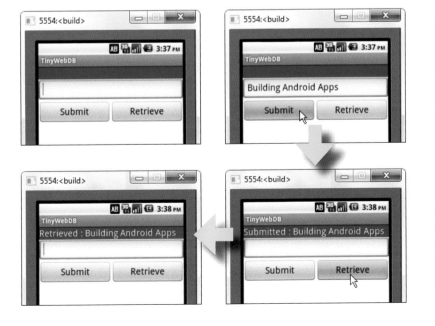

Beware

The default web service for TinyWebDB is at **http://appinvtinywebdb. appspot.com**. It is used here for demonstration but is shared by all App Inventor developers so your data will eventually be overwritten. For actual apps you should create a custom web service – as described at **appinventorbeta.com/ learn/reference/other/ tinywebdb.html**

Summary

- The LocationSensor component detects the phone's location using a combination of Wi-Fi, network, and GPS techniques

- A LocationSensor's **LocationChanged** method returns the phone's coordinates, and its **CurrentAddress** property extracts the current street location from those returned coordinates

- The OrientationSensor component detects the phone's current directional and rotational orientation

- An OrientationSensor's **OrientationChanged** method returns values expressed in degrees for Azimuth, Pitch, and Roll

- The AccelerometerSensor detects the phone's movement and calls its **Shaking** method when the phone is moved vigorously

- The EmailPicker component is simply a TextBox that automatically completes email addresses from the contact list

- The ContactPicker component's **AfterPicking** method stores the contact details in its **ContactName** and **EmailAddress** properties

- The PhoneNumberPicker's **AfterPicking** method stores the contact details in its **ContactName** and **PhoneNumber** properties

- The PhoneNumber component's **MakePhoneCall** method dials the number assigned to its **PhoneNumber** property

- The Texting component can listen for incoming messages when its **ReceivingEnabled** property is set to **true**

- A Texting component's **SendMessage** method sends its **Message** property to the number in its **PhoneNumber** property

- The Twitter component's **IsAuthorized** method fires when access is gained to a Twitter account, whereon its **SetStatus** method can be called to update that account's status

- The TinyWebDB component has a **StoreValue** method to store data with a chosen tag name, and a **GetValue** method to retrieve stored data by specifying its given tag name

9 Deploying apps

This chapter describes the creation of an Android app from initial planning through to distribution.

Planning the program

When creating a new application it is useful to spend some time planning its design. Clearly define the program's precise purpose, decide what application functionality will be required, then decide what interface components will be needed.

A plan for a simple application to pick numbers for a lottery entry might look like this:

Program purpose

- The program will generate a series of six different random numbers in the range 1 – 49, and have the ability to be reset

Functionality required

- A routine to generate and display six different random numbers

- A routine to clear the last series from display

Components needed

- Six Label components to display the series of numbers – one number per Label

- One Button component to generate and display the numbers in the Label components when this Button is clicked. This Button will not be enabled when numbers are on display

- One Button component to clear the numbers on display in the Label components when this Button is clicked. This Button will not be enabled when no numbers are on display

- One Image component to display a static image – just to enhance the appearance of the interface

Having established a program plan means you can now create the application basics by adding the components needed in Designer.

1 Open App Inventor and start a new project named "Lotto"

Lotto.apk

New App Inventor for Android Project...

Project name: Lotto

Cancel   OK

2 In the Designer, add six Label components to the Viewer from the Basic palette

3 Now add two Buttons and an Image component

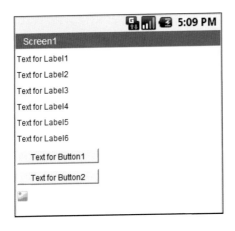

Don't forget

The Screen, Label, and Button components initially display the illustrated default text – new text properties will be assigned in the next stage described overleaf.

Assigning static properties

Having created the application basics, on the previous page, you can now assign static values using the Properties column.

Lotto.apk (continued)

1 Click anywhere on the Screen to select it then, in the Properties column, set the Screen's Title property to "Lotto Number Picker"

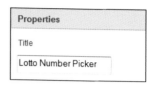

Properties

Title

Lotto Number Picker

2 Select the Button1 component then, in the Components column rename it to **btnPick** and in the Properties column set its Text property to "Get My Lucky Numbers"

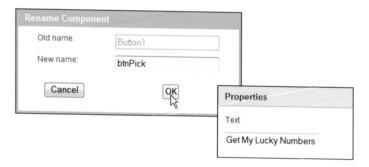

Rename Component

Old name: Button1

New name: btnPick

Cancel   OK

Properties

Text

Get My Lucky Numbers

Hot tip

The Label components in this program will have their Text property values assigned dynamically at runtime – no static properties are required.

3 Select the Button2 component then, in the Components column rename it to **btnReset** and in the Properties column set its Text property to "Reset"

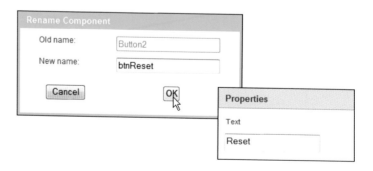

Rename Component

Old name: Button2

New name: btnReset

Cancel   OK

Properties

Text

Reset

4 Select the Image1 component then, in the Properties column, click the Picture property and click the Add button to launch the Upload File dialog

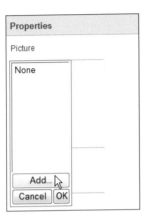

5 Now click the dialog's Browse button to select an image file to upload for addition to the project as Media resource – this action also assigns it to the Image1 Picture property

Don't forget

The project's Media resources are listed below the Components palette – the image file name gets added here.

Designing the interface

Having assigned static property values, on the previous page, you can now design the interface layout.

Lotto.apk (continued)

① Drag a HorizontalArrangement component, from the Screen Arrangement palette, and drop it at the top of the Designer Viewer – above all other components

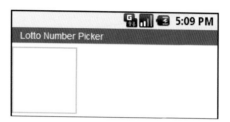

② Select the HorizontalArrangement component then set its Width property in the Properties column to "Fill parent..." – so it fills the available screen width

③ Now drag the Image1 component and drop it into the HorizontalArrangement1 component

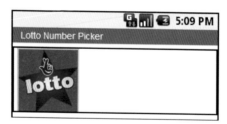

 Hot tip

You can set Width properties of the second Horizontal Arrangement and **btnPick** components to "Fill parent..." so they fill the available screen width – as shown here.

④ Next add a second HorizontalArrangement component below the first one – then drop in the Button components

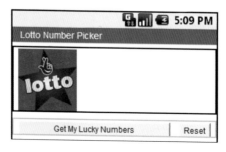

160

5 Add a TableArrangement component into HorizontalArrangement1, to the right of the Image, then set its Width and Height properties to "Fill parent...", its Columns property to 7 and its Rows property to 3

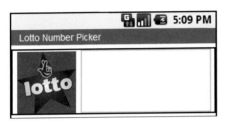

Properties
Columns
7
Rows
3
Visible
☑
Width
Fill parent..
Height
Fill parent..

6 Drop new Label components into the first cell of each table row naming them **padTop, padMid,** and **padBtm** respectively then remove their default text values

7 Set the **padTop** and **padBtm** Height property to 30 pixels and set the **padMid** Width property to 20 pixels – to center the middle row and inset it from the Image edge

Don't forget

The padding technique is useful in Table Arrangement otherwise cell contents appear in its top left corner.

8 Edit the Text properties of the original Labels by removing "Text For Label" from each and changing their FontSize to 30, then drop them one by one into the remaining cells of the middle table row

Initializing dynamic properties

Having designed the interface, on the previous page, you can now add some functionality to dynamically set the initial Text properties of the Label components and the initial Button states.

Lotto.apk (continued)

1 Launch the Blocks Editor then drag a procedure Definition block onto the workspace and name it "Clear"

2 Next snap statements into the procedure to assign an ellipsis "..." value to each Label component in turn

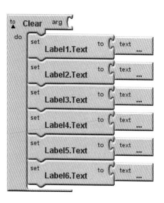

3 Now snap in further statements to specify the **Enabled** state of each Button component

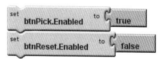

Hot tip

The initial value assigned to each Label contains a trailing space to separate them when displayed.

4 Finally add an event-handler for the screen's **Initialize** event to call the Clear procedure when the app starts

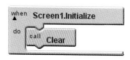

The initial state of the app is determined by the blocks on the opposite page and it is desirable to allow the user to resume the initial state by resetting the app.

 Add an event-handler for the Reset button's Click event to call the Clear procedure when the button gets tapped

This app will use three variables to store numeric values when its Pick button gets tapped so these may now be initially defined.

6 Drag a variable Definition block onto the workspace and name it "nums", then snap in a List block to create an empty list – this will store a list of randomized numbers

7 Next add a variable Definition block and name it "rand", then assign it an initial value of zero – this will store successive single random numbers as the app proceeds

8 Now add a variable Definition block and name it "temp", then assign it an initial value of zero – this will temporarily store selected single random numbers as they get shuffled to ensure that no numbers are duplicated

Adding runtime functionality

Having created blocks to initialize dynamic properties, on the previous page, you can now add procedures to provide runtime functionality to respond when the Pick button gets tapped.

Lotto.apk (continued)

1 In the Blocks Editor, drag a procedure Definition block onto the workspace and name it "Populate" – this will be used to fill the empty list variable with integer values

2 Next snap a loop statement into the procedure to count from 1 to 49, incrementing by one on each iteration

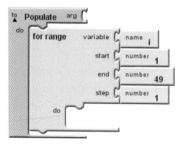

3 Now **do** snap in a statement to assign the value of the counter to the empty list variable on each iteration – filling the list with integers 1 to 49

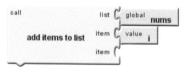

4 Drag another procedure Definition block onto the workspace and name it "Shuffle" – this will be used to rearrange the list's integer values into a random order

5 Snap a loop statement into this procedure to count from 1 to 49 again, incrementing by one on each iteration

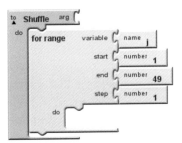

6 Now **do** snap in a statement to assign a random integer in the range 1-49 to a variable on each iteration

Don't forget

Notice that the counter variable in the second loop is named "j" to differentiate it from the first counter named "i".

7 Next **do** snap in a statement to assign the value of the list item at the counter's position to a variable on each iteration

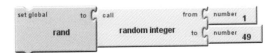

8 Finally **do** add statements to replace the list item values with others in the same range, but arranged in a random sequence and not repeating any single number

Hot tip

You don't need to understand in detail the algorithm that is used to shuffle the values.

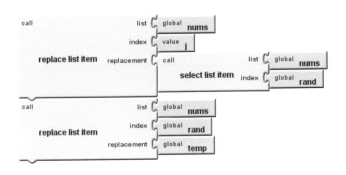

Completing functionality

Having created procedures to Populate an empty list and Shuffle the values it contains, on the previous page, you can now add a procedure to display six of the randomized list values.

Lotto.apk (continued)

1. In the Blocks Editor, drag a procedure Definition block onto the workspace and name it "Display" – this will be used to assign six list items to the Label components

2. Next snap statements into the procedure to assign the value in list item one to Label number one, the value in list item two to Label number two, and so on

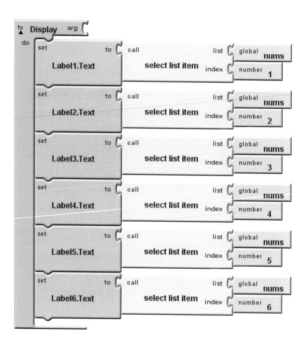

Don't forget

After the values are displayed on the Label components the Pick button becomes inactive whereas the Reset button becomes active.

3. Now snap in statements to toggle the state of the Buttons

4 Add an event-handler for the Pick button's Click event then snap in calls to the three procedures – to Populate the empty list, Shuffle its items, then Display six values

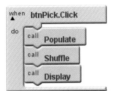

5 Run the app then tap the Buttons to see six random numbers in the range 1-49 on the Label components

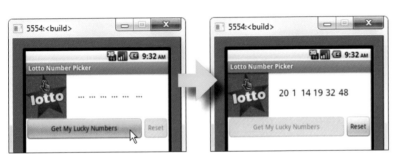

Distributing the application

Having worked through the program plan, on the previous pages, the components needed and functionality required have now been added to the app – so it's ready to be tested.

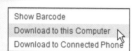

1 Click the Package For Phone in the App Inventor Designer window and download the app to your computer

2 Connect your phone to the computer and install the app on your phone

3 Launch the app to see that the Screen's Intialize event-handler has set the initial dynamic values of each Label and disabled the reset button as required

4 Now click the **btnPick** Button component to execute the instructions within its Click event-handler

A series of numbers within the desired range is displayed and the Button states have changed as required – a further series of numbers cannot be generated until the app has been reset.

5 Make a note of the numbers generated in this first series for comparison later

6 Click the **btnReset** component to execute the instructions within that Click event-handler and see the app return to its initial start-up appearance as required

7 Click the **btnPick** Button component again to execute its Click event-handler code a second time

Another series of numbers within the desired range is displayed and are different to those in the first series when compared – good, the numbers are being randomized as required.

8 Now restart the application and click the **btnPick** Button component once more

The generated numbers in this first series of numbers are different to those noted in the first series the last time the application ran – great, the random number generator is not repeating the same sequence of number series each time the application runs.

Don't forget

You can adjust the font size, Button size, and Label spacing, then test the app on several devices before distributing the app.

Selling your app

Having tested your app, as described on the previous pages, it can be distributed by forwarding the .apk file that gets downloaded using the Package For Phone, Download to this Computer option in the App Inventor Designer window. Other Android users can then install the app onto their phone, just as you did for testing.

Your app may also be sold directly to users of other Android devices via the Google Play Shop and via a number of third party "app stores". Sellers of apps on Google Play receive 70% of the selling price from the Google Checkout service with the remainder retained by carriers authorized to receive a fee for apps purchased through their network.

Submission of apps to Google Play is a relatively straightforward process that is lightly regulated:

1. Open a web browser and visit the Google Play website at **play.google.com/apps/publish/signup**

2. Complete the Listing Details giving your name and contact information to create your Developer Profile

Hot tip

Not all countries are allowed to distribute Android paid apps due to Google Checkout restrictions but the list of allowed countries is constantly growing. The latest list can be found at **http://support.google. com/googleplay/android-developer**.

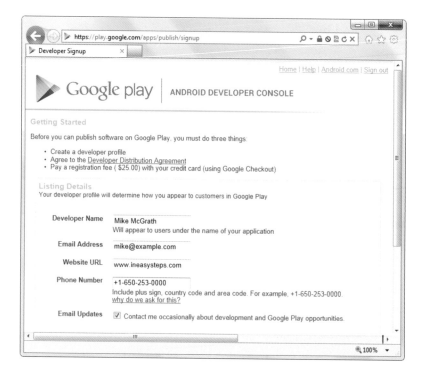

3 Pay the one-off $25 registration fee that enables you to publish software in the Google Play Shop

4 Next decide whether your app will be Free or set a price for your app in the range $0.99-$200 (0.50-100 GBP)

5 Now choose "Upload Applications" in the Developer Console then provide the Upload Assets information:

- **Package** – the .apk file name of your app (maximum 50Mb)

- **Screenshots** – the look of your app (at least two are required)

- **High Resolution Application Icon** – a required high fidelity version of the launcher icon for your app (512x512 32-bit PNG)

- **Promotional Graphic** – to advertise your app in various locations on Google Play (180w x 120h 24-bit PNG)

- **Feature Graphic** – to advertise your app in the Featured section on Google Play (1024w x 500h 24-bit PNG)

- **Video Link** – to advertise your app on YouTube

6 Finally provide the Application Information to make your app available in the Google Play Shop:

- **Language** – the language of your app (default is US English)

- **Title** – the name of your app as you would like it to appear on Google Play (one name is allowed per language)

- **Description** – the description of your app as you would like it to appear on Google Play (maximum 4000 characters)

- **Application Type** – the type of your app (Application or Game)

- **Category** – the category group for your app (such as a News & Weather Application, or an Arcade & Action Game)

- **Contact Information** – the user support channel for your app as Website, Email, or Phone (at least one is required)

Beware

Choose your app package name wisely – it must be unique and is permanent so it can't be changed later.

Hot tip

The average price of paid Android apps is around $3.00 – similar to the price of paid Apple apps.

171

Summary

- A program plan should define the app's precise purpose, functionality required, and components needed

- The static properties of components, that DO NOT change when the app runs, can be set in their Properties column

- Images that are required to appear in the interface need to be uploaded for the app to use as a Media resource

- HorizontalArrangement, VerticalArrangement, and Table-Arrangement components are useful for interface layout

- Label components with no text can be used for layout padding by specifying explicit Width and Height property values

- The dynamic properties of components, that DO change when the app runs, can be set to an initial value in their Properties column – or by the Screen's Initialize event-handler

- Changing the value of a component's Enabled property toggles its state between active and inactive

- Procedures can provide runtime functionality to respond to user actions in the interface

- The result provided when a runtime function is called can be displayed on a Label component in response to a user action

- An app should be tested on several devices before distribution to ensure it performs as expected

- Complete tested apps may be sold directly to users of other Android devices via the Android Marketplace

- Submission of apps to the Android Marketplace is a straightforward process that simply requires payment of a registration fee and upload of the app assets

Handy reference

This chapter describes the Built-in blocks you can use to build Android apps with App Inventor development.

Definition blocks

procedure

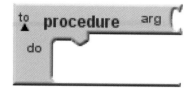

The **procedure** block groups together a series of statements that get executed whenever the procedure is called. Optionally arguments for the procedure may be specified using **name** blocks. When a procedure is created App Inventor automatically provides a unique name and generates a call block for that procedure in the My Definitions drawer. Procedures can be renamed by editing the block's label but the name must be unique in the app. App Inventor automatically renames associated call blocks to match.

procedureWithResult

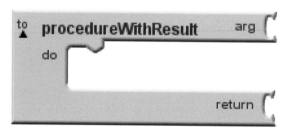

The **procedureWithResult** block is like the **procedure** block but additionally returns a result to the caller after executing its statements.

| [ignore result]

The | block provides a socket to attach a block where there is otherwise no socket available. For example, within the **do** part of a procedure block. The attached block will be executed but its returned result will be ignored.

variable

The **variable** block creates a named container to store a value that can be changed while the app is running. Variables are "global" in scope – so they can be referenced from any code in the app. When a variable is created App Inventor automatically creates two associated blocks in the My Definitions drawer:

· The **global** block retrieves the value of the variable.

· The **set global** block assigns a value to the variable.

App Inventor automatically gives new variables a unique name. Variables can be renamed by editing the block's label but the name must be unique within that app. App Inventor will automatically rename the associated blocks to match.

name

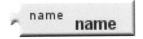

The **name** block creates a named argument that can be used when calling a procedure by snapping it into the procedure's **arg** socket.

For each **name** block that is created App Inventor creates an associated **value** block and places it in the My Definitions drawer. That **value** block refers to the value of the argument that is passed to the procedure when the procedure is called.

App Inventor automatically gives new arguments a unique name. Arguments can be renamed by editing the block's label in an app but the name must be unique within that app. App Inventor will automatically rename the associated blocks to match.

Text blocks

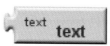

The **text** block contains a string of text.

The **join** block appends the second specified string to the first, whereas the **make text** block unites all strings into one string.

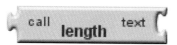

The **length** block returns the character length of the specified string.

The **is text empty?** block returns **true** when the specified string is empty, otherwise it returns **false**.

The **text=** block returns **true** only when the specified strings are identical, having the same characters in the same order.
The **text<** block returns **true** when the first string is alphabetically less than the second string, whereas the **text>** block returns **true** when the first string is alphabetically greater than the second string.

The **trim** block returns a copy of the specified string with all leading and trailing spaces removed.

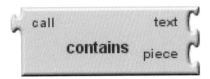

The **upcase** block returns a copy of the specified string converted to uppercase, whereas the **downcase** block returns a copy of the specified string converted to lowercase.

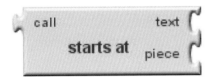

The **contains** block returns **true** when the specified piece is found in the specified string, otherwise it returns **false**.

The **starts at** block returns the character position where the first character of the **piece** first appears in the **text** string, or 0 if absent.

The **split** block divides a string into pieces using **at** as the dividing point. Similar divisions can be made with the **split at first**, **split at first of any**, **split at any**, and **split at spaces** blocks.

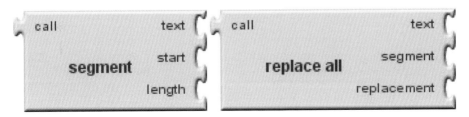

The **segment** block returns a substring from the **start** position and of the specified **length**, whereas the **replace all** block returns a substring in which the specified **replacement** has been substituted.

Lists blocks

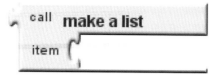

The **make a list** block creates a list from the specified **item** blocks.

The **select list item** block selects the item at the specified **index** position in the specified list – the first list item is at position 1.

The **replace list item** block substitutes the specified **replacement** into the specified **list** at the specified **index** position.

The **remove list item** block removes the item from the specified **list** at the specified **index** position.

The **is a list?** block returns **true** if the specified **thing** is a list, whereas the **is list empty?** block returns **true** if the list has no items.

The **length of list** block returns the number of items in the list, and the **pick random item** block returns a randomly selected item.

The **append to list** block appends the second list to the first list, whereas the **add items to list** block appends single items to the list.

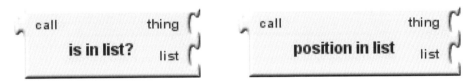

The **is in list?** block returns **true** if the specified **thing** is a list item, whereas the **position in list** block returns its index position.

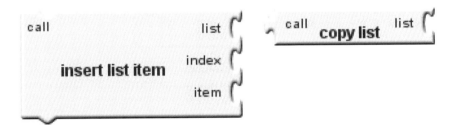

The **insert list item** block inserts a specified **item** into a specified **list** at the specified **index** position, whereas the **copy list** block returns a copy of the specified list.

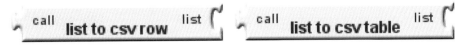

The **list to csv row** block returns the list items as a comma-separated list, whereas the **list to csv table** block returns the list items as a comma-separated list with final carriage returns.

Math blocks

The **number** block specifies a numeric value, whereas the **is a number?** block returns **true** if the specified **thing** is a number.

The arithmetic operator blocks **+**, **-**, **x**, and **/** return the result of the operation performed on two specified numeric operands.

The **modulo** block returns the remainder after dividing the first operand by the second operand with the same +/- sign as the first operand, whereas the **remainder** block returns the remainder with the same +/- sign as the second operand.

The comparison operator blocks **<**, **<=**, **>**, and **>=** return **true** if the numeric comparison is valid, otherwise they return **false**.

The equality operator blocks **=** and **not=** return **true** if the numeric comparison is valid, otherwise they return **false**.

The **min** block returns the smallest of a specified set of numbers, whereas the **max** block returns the largest number.

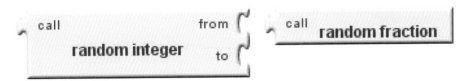

The **random integer** block returns a random integer value between two specified values, whereas the **random fraction** block returns a random value between 0 and 1.

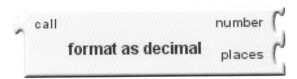

The **floor** block returns the nearest integer below a specified number, whereas the **ceiling** block returns the nearest integer above.

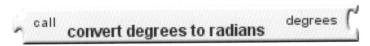

The **round** block returns the nearest integer to a specified number, whereas the **negate** block returns the negative of a specified number.

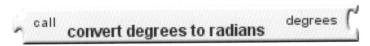

The **format as decimal** block returns the specified **number** with a specified number of **places** added after the decimal point.

The **convert degrees to radians** block returns the specified **degrees** as radians, whereas the **convert radians to degrees** block returns degrees.

The mathematic blocks **sin, cos, tan, asin, acos, atan, atan2, exp, expt, log, quotient,** and **abs** return the math equivalent of specified values.

Logic blocks

true

The **true** block represents the boolean true constant value and can be used to specify the value of a variable that describes a condition or set the boolean property of a component to make it active.

false

The **false** block represents the boolean false constant value and can be used to specify the value of a variable that describes a condition or set the boolean property of a component to make it inactive.

and

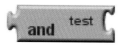

The **and** block tests whether all of a specified set of logical conditions are **true**. It returns **true** when each and every specified condition is **true**, otherwise it returns **false**.

or

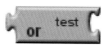

The **or** block tests whether any of a specified set of logical conditions are **true**. It returns **true** when any one specified condition is **true**, otherwise it returns **false**.

not

The **not** block performs logical negation, returning **true** if the specified condition is **false** and conversely returning **false** if the specified condition is **true**.

equals

The **equals** block tests whether its two specified values are equal and returns **true** when the test succeeds, otherwise it returns **false**.

- Two **number** blocks are equal if their values are numerically equal. For example, 1 is equal to 1.0

- Two **text** blocks are equal if their values have the same characters in the same order, and also have the same case. For example, "banana" is not equal to "Banana"

- Numbers and text are equal if the number is numerically equal to a number that would be printed with that text. For example, 12.0 is equal to the result of joining the first character of 1A to the last character of Teafor2

- Two lists are equal if they have the same number of elements and the corresponding elements are equal

Control blocks

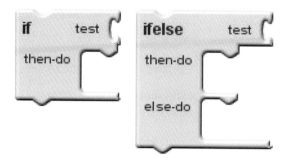

The **if** block performs a conditional test then executes the statements it contains only when the test returns a **true** value. The **ifelse** block executes its **then-do** statements when the test returns **true** but executes its **else-do** statements when the test returns **false**.

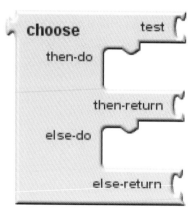

The **choose** block performs a conditional test then executes its **then-do** statements and returns its **then-return** value when the test returns **true**, otherwise it executes its **else-do** statements and returns its **else-return** value.

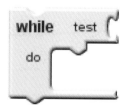

The **while** block performs a conditional test then executes its **do** statements when the test returns a **true** value, then tests again and executes its **do** statements again until the test returns **false**.

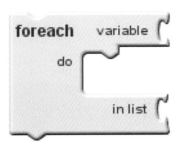
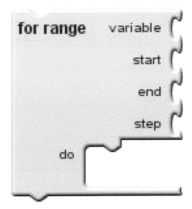

The **foreach** block executes its **do** statements once for each item in the specified list, whereas the **for range** block executes its **do** statements each time a specified variable steps through a specified numeric range.

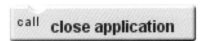

The **get start text** block returns any text that was passed to the app when it was started, such as user input.

The **close screen** block closes the app, whereas the **close screen with result** block closes the app and also assigns a specified result to a system APP_INVENTOR_RESULT variable – so its value is available to any subsequent running app.

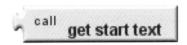

The **close application** block exits the app.

Colors blocks

The **make color** block creates a color object from a specified list comprising four items – describing that color's Red, Green, Blue and Alpha elements in the range 0-255, where 255 is full opacity.

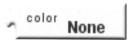

The **split color** block returns a list of four items describing the specified **color** object's Red, Green, Blue and Alpha elements in the range 0-255.

The **none** block specifies the color White with zero opacity.

Other named color blocks specify the value of common colors.

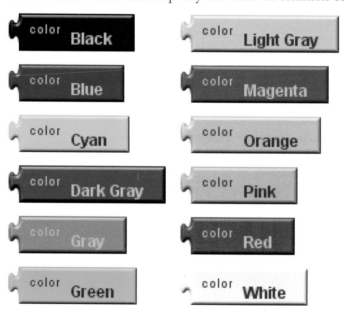

Index

189

W

Z

U

V